Passive Planar Microwave Devices

Passive Planar Microwave Devices

Editors

Armando Fernandez-Prieto
Alejandro Javier Martinez-Ros

MDPI • Basel • Beijing • Wuhan • Barcelona • Belgrade • Manchester • Tokyo • Cluj • Tianjin

Editors
Armando Fernandez-Prieto
Universidad de Sevilla
Spain

Alejandro Javier Martinez-Ros
Universidad de Sevilla
Spain

Editorial Office
MDPI
St. Alban-Anlage 66
4052 Basel, Switzerland

This is a reprint of articles from the Special Issue published online in the open access journal *Applied Sciences* (ISSN 2076-3417) (available at: https://www.mdpi.com/journal/applsci/special_issues/Passive_Planar_Microwave_Devices).

For citation purposes, cite each article independently as indicated on the article page online and as indicated below:

LastName, A.A.; LastName, B.B.; LastName, C.C. Article Title. *Journal Name* **Year**, *Volume Number*, Page Range.

ISBN 978-3-0365-4495-3 (Hbk)
ISBN 978-3-0365-4496-0 (PDF)

Contents

About the Editors

Armando Fernandez-Prieto

Armando Fernandez-Prieto was born in Ceuta, Spain, in September 1981. He received the Licenciado and Ph.D. degrees in physics from the Universidad de Sevilla, Seville, Spain, in 2007 and 2013, respectively. During 2010 he was a visiting researcher at Heriot-Watt University, Edinburgh, UK, at Prof. Jiasheng Hong's group where he was involved in the design of multilayer LCP balanced microwave filters. He was also visiting researcher in 2017 at Universidad Autónoma de Barcelona, Barcelona, Spain at CIMITEC under Prof. Ferran Martín's supervision. During this period Dr. Fernández-Prieto was involved in the design of microwave sensors, slow-wave structures and balanced-to-balanced planar diplexers. From 2018 to 2022 Dr. Fernández-Prieto was Assistant Professor of electromagnetism with the Department of Electronics and Electromagnetism, University of Seville. Nowadays, he is currently an Associate Professor of electromagnetism with the Department of Electronics and Electromagnetism, University of Seville, and still a member of the Microwaves Group. His research interests focus on printed passive microwave components, applications on filter theory to the design of microwave antennas and metamaterials. Dr. Fernández-Prieto was awarded in 2015 and 2017 with the best research paper prize from University of Seville in the areas of Engineering and Science, respectively.

Alejandro Javier Martinez-Ros

Alejandro Javier Martinez-Ros was born in Cartagena, Spain, in 1981. He received the Telecommunication Engineer and Ph.D. degrees from the Universidad Politécnica de Cartagena (UPCT), Cartagena, Spain, in 2005 and 2014, respectively. During 2006, he worked as an RF Engineer at the IMST GmbH in Kamp-Lintfort, Germany, where he was involved with the development of EBG structures for high-precision antennas for the Galileo system, and with phase array and patch antennas for global navigation satellite systems. In 2007, he moved to Spain and worked as a Research Engineer in a naval company, where he participated in the design of submarine telecommunication systems based on underwater acoustics. In 2008, he joined the Telecommunication and Electromagnetic Group (GEAT), UPCT, as a Research Assistant. He has been a visiting Ph.D. student at the Queen's University Belfast, Northern Ireland, U.K. and a Postdoctoral Researcher in the Microwave Laboratory, University of Pavia, Pavia, Italy. Currently, he is Associate Professor in the Microwave Group of the Universidad de Sevilla, Seville, Spain. His scientific research is mainly focused on the analysis and synthesis of leaky-wave devices for microwave and millimeter waveband applications. Dr. Martínez-Ros was awarded in 2015 with the best doctoral thesis prize from UPCT and the best doctoral thesis from COIT-HISDESAT. He is the coauthor of the best student paper at the 2013 URSI Symposium, finalist for the best student paper at the 2013 AP-S/URSI, recipient of a travel grant at the 2011 IEEE MTT-S IMWS, and of the second prize for the best student paper at the 2012 URSI Symposium.

applied sciences

Editorial

Passive Planar Microwave Devices

Alejandro Javier Martínez-Ros [1,*] **and Armando Fernandez-Prieto** [2,*]

[1] Department of Applied Physics I, Universidad de Sevilla, 41013 Seville, Spain
[2] Department of Electronics and Electromagnetism, Universidad de Sevilla, 41013 Seville, Spain
[*] Correspondence: amartinez49@us.es (A.J.M.-R.); armandof@us.es (A.F.-P.)

Citation: Martínez-Ros, A.J.; Fernandez-Prieto, A. Passive Planar Microwave Devices. *Appl. Sci.* **2022**, *12*, 4444. https://doi.org/10.3390/app12094444

Received: 22 April 2022
Accepted: 25 April 2022
Published: 28 April 2022

Publisher's Note: MDPI stays neutral with regard to jurisdictional claims in published maps and institutional affiliations.

1. Introduction

Passive planar circuits play a key role in many RF/microwave applications, such as in wireless communications, medical instrumentation, and remote sensing. From their earliest developments during World War II to the present day, they have become indispensable for their low cost and low weight while maintaining high performance. As a result, they are still undergoing research and development. In recent years, multiple technologies have been proposed with the aim of combining the characteristics of traditional planar and nonplanar transmission lines, highlighting substrate integrated waveguide (SIW) technology as the most popular among them.

This Special Issue is focused on highlighting recent contributions to microwave device development in planar technologies. A total of twelve papers have been published in this volume, each addressing several important research problems and advancements in the field of filters, multiport circuits, dividers, combiners, couplers, multiplexers, microwave sensors, and antennas. These articles provide a significant contribution to the state of the art.

2. Contributions

Planar technologies allow for a wide number of possibilities when facing new designs and challenges. In this Special Issue, a compilation of twelve relevant papers on the topic of passive planar microwave devices has been published.

In [1], ElKhorassani et al. presented the design of a two-port electronically tunable phase shifter with progressive impedance transformation in the K band. The phase shifter consists of a 3 dB hybrid coupler loaded with reflective phase-controllable circuits. They measured the prototype at a frequency of 18 GHz, showing a dynamic range of 600 degrees. Zhu et al. [2] proposed a new method to monitor patients with Huntington's disease (HD) using wireless sensing technology. Experimental data were firstly collected by the self-developed microwave sensing platform (MSP) and subsequently preprocessed. Finally, support vector machine (SVM) and random forest (RF) algorithms were used to train the model. The MSP system continuously monitors patients in a noncontact manner, which offers more convenience and privacy. The experimental results show that the prediction accuracy of SVM can be as high as 98.0%, and that of RF can reach 96.7%. Camacho-Gomez et al. [3] proposed a design for a multiband textile antenna for LTE and 5G communication services, composed of a rectangular microstrip patch, two concentric annular slots, and a U-shaped slot. They showed that the coral reef optimization, with the substrate layer (CRO-SL) algorithm, can produce a robust multiband textile antenna, including the LTE and 5G frequency bands. For the optimization process, the CRO-SL is guided by means of a fitness function obtained after antenna simulation by a specific simulation software for electromagnetic analysis in the high-frequency range. In [4], Kumar et al. presented a planar, microstrip line-fed, quad-port, multiple-input multiple-output (MIMO) antenna with dual-band rejection features for ultra-wideband (UWB) applications. The proposed MIMO antenna design consists of four identical octagonal-shaped radiating elements that are placed orthogonally to each other. The dual-band rejection property (3.5 GHz and 5.5 GHz corresponding to Wi-MAX and WLAN bands) was obtained by introducing a hexagonal-shaped complementary split-ring resonator (HCSRR) in the radiators of the designed antenna. Isolation was observed to be higher than 18 dB, and the envelope correlation coefficient (ECC) was less than 0.07 for the MIMO/diversity antenna in the operating range of 3–16 GHz. Dai et al. [5] proposed an ultra-wideband and miniaturized spoof plasmonic antipodal Vivaldi antenna (AVA) for ultra-wideband antenna applications, such as in the 2G/3G/4G/5G base stations. They designed and measured an antenna operating from 1.8 GHz to 6 GHz, with an average gain of 7.24 dBi.

Muñoz-Enano et al's [6] review paper focuses on microwave sensors, particularly planar sensors based on resonant elements for material characterization, including solids and liquids. In addition, they reported the main advantages and limitations of microwave sensors based on electrically small planar resonant elements. Moreover, they presented several sensing strategies, with a special emphasis on differential-mode sensors. In [7], Medran del Rio et al. proposed a new dual-band balanced bandpass filter based on magnetically coupled open-loop resonators in multilayer technology. The lower differential passband, centered at the global positioning system (GPS) L1 frequency, 1.575 GHz, was created by means of two coupled resonators etched in the middle layer of the structure, while the upper differential passband, centered at a Wi-Fi frequency of 2.4 GHz, was generated by coupling two resonators on the top layer. Magnetic coupling was used to design both passbands, leading to an intrinsic common-mode rejection of 39 dB within the lower passband and 33 dB within the upper passband. Arnberg et al. [8] demonstrated the beneficial effects of introducing glide symmetry into a two-dimensional periodic structure. Specifically, they investigated dielectric parallel plate waveguides periodically loaded with Jerusalem cross slots in three configurations: conventional, mirror, and glide symmetry. Out of these three configurations, they demonstrated that the glide-symmetric structure is the least dispersive and has the most isotropic response. Furthermore, the glide-symmetric structure achieved the highest effective refractive index, which enables the realization of a broader range of electromagnetic devices. To illustrate the potential of this glide-symmetric unit cell, a Maxwell fish-eye lens was designed to operate at 5 GHz.

Trujillo-Flores et al. [9] developed a fully transparent multiband antenna for vehicle communications. The antenna is coplanar-waveguide-fed to facilitate its manufacture and increase its transmittance. An indium–tin–oxide film, a type of transparent conducting oxide, was selected as the conductive material for the radiation path and ground plane, with a sheet resistance of 8 ohms/square. The substrate is glass with a relative permittivity of 5.5. The proposed antenna meets the frequency requirements for vehicular communications according to the IEEE 802.11p standards. Additionally, it covers the frequency bands from 1.82 to 2.5 GHz for LTE communications applied to vehicular networks. Cui et al. [10] proposed a multilayer bandpass filter with high selectivity. To this aim, the discriminating coupling formed by slot-coupled quarter-wavelength and half-wavelength resonators introduced a zero at $3f_0$ (f_0 is the center frequency), while the second harmonic was also suppressed due to the quarter-wavelength resonators. As a result of the multilayer structure, source–load coupling was introduced to improve selectivity. Then, an extra coupled line path was added with the same amplitude as the discriminating coupling path while they were out of phase. Thus, signal cancelation produced three extra transmission zeros, with a further improvement in selectivity and suppression performance. The design was validated by a manufactured prototype bandpass filter centered at 2.49 GHz with a fractional 3 dB bandwidth of 8.1%. Kim et al. [11] designed a microwave planar cutoff probe to measure the cutoff frequency in a transmission (S21) spectrum. For real-time electron density measurement in plasma processing, three distinct types of probes were demonstrated: point-type, ring-type (RCP), and bar-type (BCP). Moreover, the work includes a computational characterization of an RCP and a BCP with various geometrical parameters, as well as a plasma parameter, through a commercial three-dimensional electromagnetic simulation. Among the investigated parameters, the antenna distance was found to be the most important parameter for improving the accuracy of both RCP and BCP. Finally,

Herraiz et al. [12] proposed a new wideband transition from the microstrip line (MS) to ridge empty substrate integrated waveguide (RESIW), with a dielectric taper based on the equations of the superellipse. The new wideband transition presents simulated return losses in a back-to-back transition greater than 20 dB in an 87% fractional bandwidth, while in the previous transition, the fractional bandwidth was 82%, which supposes an increase of 5%. Furthermore, the transition presented simulated return losses greater than 26 dB in an 84% fractional bandwidth, while the measured return loss was above 14 dB with an insertion loss lower than 1 dB throughout the band.

Author Contributions: Conceptualization, A.J.M.-R. and A.F.-P.; Writing—original draft preparation, A.J.M.-R. and A.F.-P.; writing—review and editing, A.J.M.-R. and A.F.-P. All authors have read and agreed to the published version of the manuscript.

Funding: This work was supported by Grant PID2020-116739GB-I00 funded by MCIN/AEI/10.13039/501100011033.

Conflicts of Interest: The authors declare no conflict of interest.

References

1. ElKhorassani, M.; Palomares-Caballero, A.; Alex-Amor, A.; Segura-Gómez, C.; Escobedo, P.; Valenzuela-Valdés, J.; Padilla, P. Electronically Controllable Phase Shifter with Progressive Impedance Transformation at K Band. *Appl. Sci.* **2019**, *9*, 5229. [CrossRef]
2. Zhu, Q.; Guan, L.; Khan, M.; Yang, X. Monitoring of Huntington's Disease Based on Wireless Sensing Technology. *Appl. Sci.* **2020**, *10*, 870. [CrossRef]
3. Camacho-Gomez, C.; Sanchez-Montero, R.; Martínez-Villanueva, D.; López-Espí, P.; Salcedo-Sanz, S. Design of a Multi-Band Microstrip Textile Patch Antenna for LTE and 5G Services with the CRO-SL Ensemble. *Appl. Sci.* **2020**, *10*, 1168. [CrossRef]
4. Kumar, P.; Urooj, S.; Alrowais, F. Design and Implementation of Quad-Port MIMO Antenna with Dual-Band Elimination Characteristics for Ultra-Wideband Applications. *Appl. Sci.* **2020**, *10*, 1715. [CrossRef]
5. Dai, L.; Tan, C.; Zhou, Y. Ultrawideband Low-Profile and Miniaturized Spoof Plasmonic Vivaldi Antenna for Base Station. *Appl. Sci.* **2020**, *10*, 2429. [CrossRef]
6. Muñoz-Enano, J.; Vélez, P.; Gil, M.; Martín, F. Planar Microwave Resonant Sensors: A Review and Recent Developments. *Appl. Sci.* **2020**, *10*, 2615. [CrossRef]
7. Medran del Rio, J.; Lujambio, A.; Fernández-Prieto, A.; Martinez-Ros, A.; Martel, J.; Medina, F. Multilayered Balanced Dual-Band Bandpass Filter Based on Magnetically Coupled Open-Loop Resonators with Intrinsic Common-Mode Rejection. *Appl. Sci.* **2020**, *10*, 3113. [CrossRef]
8. Arnberg, P.; Barreira Petersson, O.; Zetterstrom, O.; Ghasemifard, F.; Quevedo-Teruel, O. High Refractive Index Electromagnetic Devices in Printed Technology Based on Glide-Symmetric Periodic Structures. *Appl. Sci.* **2020**, *10*, 3216. [CrossRef]
9. Trujillo-Flores, J.; Torrealba-Meléndez, R.; Muñoz-Pacheco, J.; Vásquez-Agustín, M.; Tamariz-Flores, E.; Colín-Beltrán, E.; López-López, M. CPW-Fed Transparent Antenna for Vehicle Communications. *Appl. Sci.* **2020**, *10*, 6001. [CrossRef]
10. Cui, J.; Chang, H.; Zhang, R. High Selectivity Slot-Coupled Bandpass Filter Using Discriminating Coupling and Source-Load Coupling. *Appl. Sci.* **2020**, *10*, 6807. [CrossRef]
11. Kim, S.; Lee, J.; Lee, Y.; Yeom, H.; Lee, H.; Kim, J.; You, S. Computational Characterization of Microwave Planar Cutoff Probes for Non-Invasive Electron Density Measurement in Low-Temperature Plasma: Ring- and Bar-Type Cutoff Probes. *Appl. Sci.* **2020**, *10*, 7066. [CrossRef]
12. Herraiz, D.; Esteban, H.; Martínez, J.; Belenguer, A.; Cogollos, S.; Nova, V.; Boria, V. Transition from Microstrip Line to Ridge Empty Substrate Integrated Waveguide Based on the Equations of the Super ellipse. *Appl. Sci.* **2020**, *10*, 8101. [CrossRef]

Communication

Electronically Controllable Phase Shifter with Progressive Impedance Transformation at K Band

Mohamed T. ElKhorassani [1], Angel Palomares-Caballero [1,*], Antonio Alex-Amor [1,2], Cleofás Segura-Gómez [1], Pablo Escobedo [3], Juan F. Valenzuela-Valdés [1] and Pablo Padilla [1]

[1] Department of Signal Theory, Telematics and Communications, Universidad de Granada—CITIC, 18071 Granada, Spain; taha.elkhorasssani@gmail.com (M.T.E.); aalex@gr.ssr.upm.es (A.A.-A.); cleofas@correo.ugr.es (C.S.-G.); juanvalenzuela@ugr.es (J.F.V.-V.); pablopadilla@ugr.es (P.P.)
[2] Information Processing and Telecommunications Center, Universidad Politécnica de Madrid, 28040 Madrid, Spain
[3] Bendable Electronics and Sensing Technologies (BEST) Group, School of Engineering, University of Glasgow, Glasgow G12 8QQ, UK; pablo.escobedo@glasgow.ac.uk
* Correspondence: angelpc@ugr.es

Received: 11 November 2019; Accepted: 29 November 2019; Published: 1 December 2019

Abstract: This communication presents the design of a two-port electronically tunable phase shifter at K band. The phase shifter consists of a 3 dB hybrid coupler loaded with reflective phase-controllable circuits. The reflective circuits are formed by varactors and non-sequential impedance transformers which increase the operational bandwidth and the provided phase shift. The final phase shifter design is formed by two loaded-coupler stages of phase shifting to guarantee a complete phase turn. An 18 GHz phase shifter design with dynamic range of 600 degrees of phase shift is depicted in this document. The prototype is manufactured and validated through measurements showing good agreement with the simulation results.

Keywords: microwave tunable phase shifters; 3 dB/90° coupler; K band

---

1. Introduction

Antenna arrays have been one of the most used structural strategies for the design of directive antennas in microwave ranges with demanding radiation requirements [1]. In the last years, planar substrate-based antennas have been present in many phased array designs due to the low manufacturing costs, easy integration, and low weight and profile [2,3]. This is particularly of interest in array structures such as reflectarrays and transmitarrays. In the case of the reflectarrays, the conformed reflectors are replaced by planar array structures that introduce at each array cell the desired phase shift, whose value is the one provided by the corresponding point of the equivalent reflector [4]. The transmitarrays are based on the same working principle, but the structure replaced by the periodic array is a lens instead of a reflector [5]. Both options have become a real alternative in reception/transmission systems whose specifications imply demanding requirements in terms of matching, directivity, gain, radiation pattern, or aperture efficiency [6]. The phase shifting strategy for both reflectarray- and transmitarray-based antennas depends on the system architecture and is typically obtained by a phase-delay circuit at each radiating cell of the array [7,8]. This phase shift can be fixed in design, yielding a passive device, or tunable, yielding an active antenna. The tunable nature of the phase shifter can be provided either mechanically or electronically [6,9]. Many of the designs of electronically controllable phase shifters are based on reflective inductance-capacitance (LC) tunable circuits [10], most of them focused on frequencies up to 3 GHz [11–14]. This latter approach is the one that guides this work, leading to the design of electronically reconfigurable phase shifters at higher microwave ranges [10]. In this paper we present a K band electronically tunable phase shifter based on reflective circuits with progressive impedance transformers and varactors. These impedance transformers provide an enhanced frequency bandwidth and an increase in the phase variation produced in the phase shifter [15].

The relevance of our work is based on the combination of three main aspects, which are: the use of impedance transformers for increasing the phase variation range, the unitary phase shifter cascading, and the high operating frequency (18 GHz). These three elements combined confer interest to our work and provide a significant contribution related to previous works, such as [10] and [15]. In particular, the idea of using of impedance transformers proposed in [15] is now validated through prototypes. The document is organized as follows: Section 2 depicts the phase shifter working principle and presents the shifter design, while Section 3 presents the validation through the manufacture and measurement of a prototype. Finally, conclusions are outlined in Section 4.

2. Materials and Methods: Phase Shifter Working Principle and Design

The working principle of the device is based on the combination of two microwave elements: a 3 dB/90 hybrid coupler and a couple of reflective tunable circuits. The reflective circuits are connected to the −3 dB/90° and −3 dB/180° output ports (ports 3 and 4, respectively), while the isolated port of the 3 dB/90° coupler becomes the output port (port 2). Figure 1 shows the connection scheme. The input signal (port 1) is conducted through the coupler to ports 3 and 4 with a phase difference of 90° between them. The signal reflections in ports 3 and 4 steer both signals again towards the coupler, adding the reflective circuit tunable phase variation ($\Delta\varphi$). As a result, port 1 receives no signal (opposite phase summation of identical signals) and all the available power is routed to port 2 with a phase shift of $\Delta\varphi$. For the reflective circuits, we propose the use of a progressive impedance transformation, for enlarging the achievable phase range.

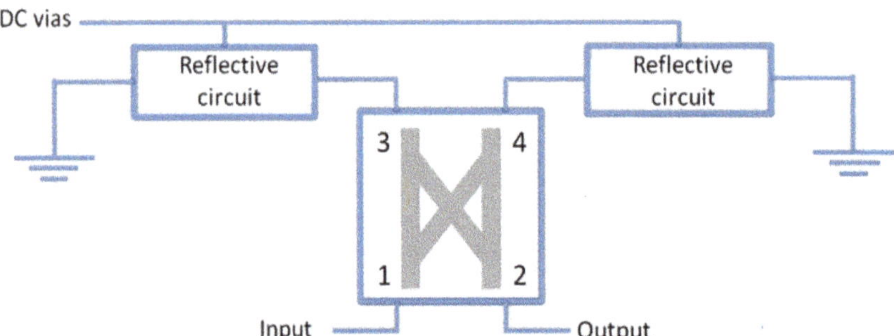

Figure 1. Scheme of the reflection-based phase shifter.

2.1. Baseline 3 dB/90° Coupler

The basic element of the phase shifter is a conventional 3 dB/90° coupler. The coupler is designed for a working frequency of 18 GHz with a bandwidth higher than 2 GHz considering −20dB in the $|S_{11}|$ level. The coupler symmetry imposes identical amplitude values levels at ports 3 and 4, with an accurate 90° phase difference between these ports. Figure 2 provides the model of the microstrip coupler, along with its performance results.

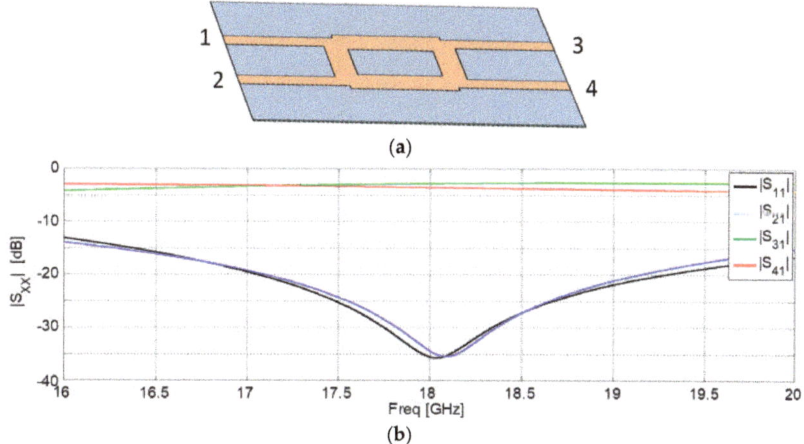

Figure 2. Three decibel hybrid coupler: (**a**) microstrip design; (**b**) design performance ($|S_{XX}|$ values).

2.2. Reflective LC Circuit with Progressive Impedance Transformation

The key element of the phase shifter that provides its reconfigurability is a reflective circuit. It is composed of a LC circuit with a short-circuit to the ground plane at the end. The capacitive effect is provided by a varactor while the inductive effect is provided by narrowing a section in the microstrip line (high impedance section). The varactor used for this design is the MA46585 varactor, with a capacity range of 0.13 pF to 2.2 pF. The connection to the ground plane is made with a metallic via. Due to the low resistive value of the varactors (~9 Ω), it is necessary to add a progressive impedance transformation to the reflective circuit, in order to improve the performance of the reflective element. For the 9-to-50 Ω transformation an impedance conversion based on two cascaded quarter-wavelength transformers is designed. The sequential transformation of the impedance in two steps impedance (i.e., transformation of 50 Ω to 25 Ω and 25 Ω to 9 Ω) requires higher microstrip widths. These widths are impracticable in a size-reduced device. The alternative is a non-sequential transformation in two steps: to avoid low line impedances, a new relation between impedances can be set up, according to Equation (1). This principle of impedance transformation is illustrated in Figure 3, where Z_x are the characteristic impedances of the microstrip sections.

$$Z_{in} = \frac{Z_0^2}{Z_x} = \frac{Z_0^2}{Z_1^2/Z_0} = \frac{Z_0^3}{Z_1^2} \tag{1}$$

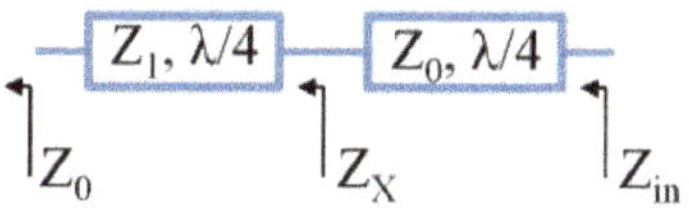

Figure 3. Non-sequential cascaded $\lambda/4$ impedance transformers according to Equation (1).

The non-sequential transformation requires a quarter-wavelength section of impedance Z_1 of 118 Ω, which means a thinner microstrip section. Therefore, two cascaded quarter-wave impedance transformers with characteristic impedance of 118 Ω and 50 Ω, respectively, are used to transform the circuit impedance from 50 Ω to 9 Ω along a width bandwidth. Figure 4 provides the circuit model of the non sequential transformer, and its design outcomes.

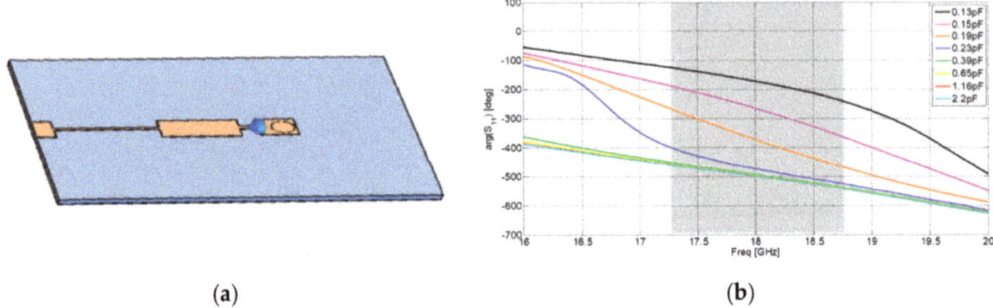

(a)   (b)

Figure 4. Design results of the reflective LC circuit with progressive impedance transformation: (**a**) simulation design; (**b**) phase shift produced.

The total phase variation is of around 290° for the complete capacity range [0.13–2.2 pF], but with a non-linear behavior. However, this non-linearity is compensated by the DC-voltage/capacity relation. Compared to the case without progressive impedance transformation, the latter provides a phase shift of around 220°. This means an improvement of at least 70° due to the progressive impedance transformation, for the same capacity range.

2.3. Complete Phase Shifter Design and Performance

The phase shift provided by a loaded coupler is not enough to guarantee a phase variation higher than 360°. Therefore, in the complete device, two cascaded sets are connected. Figure 5 provides the complete phase shifter model, together with its performance results. The desired band in the design is at least of 1.5 GHz, centered at 18 GHz.

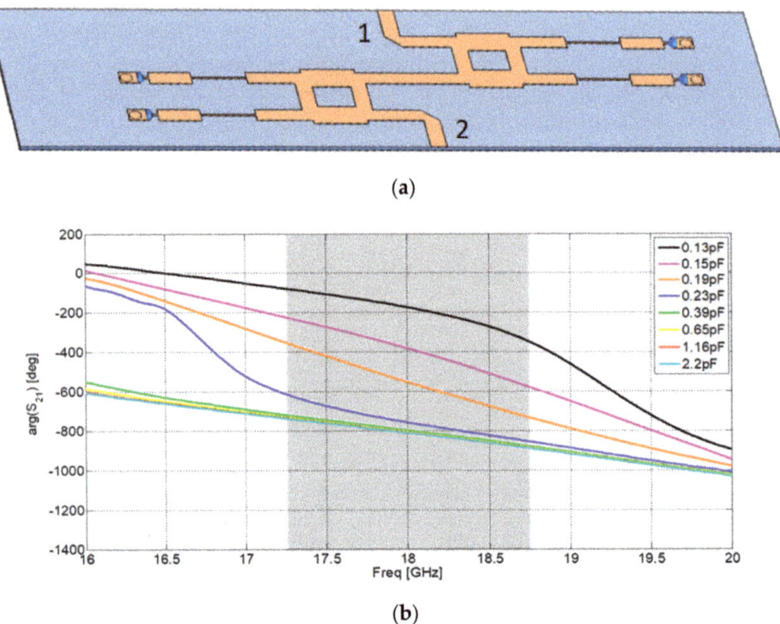

(a)

(b)

Figure 5. *Cont.*

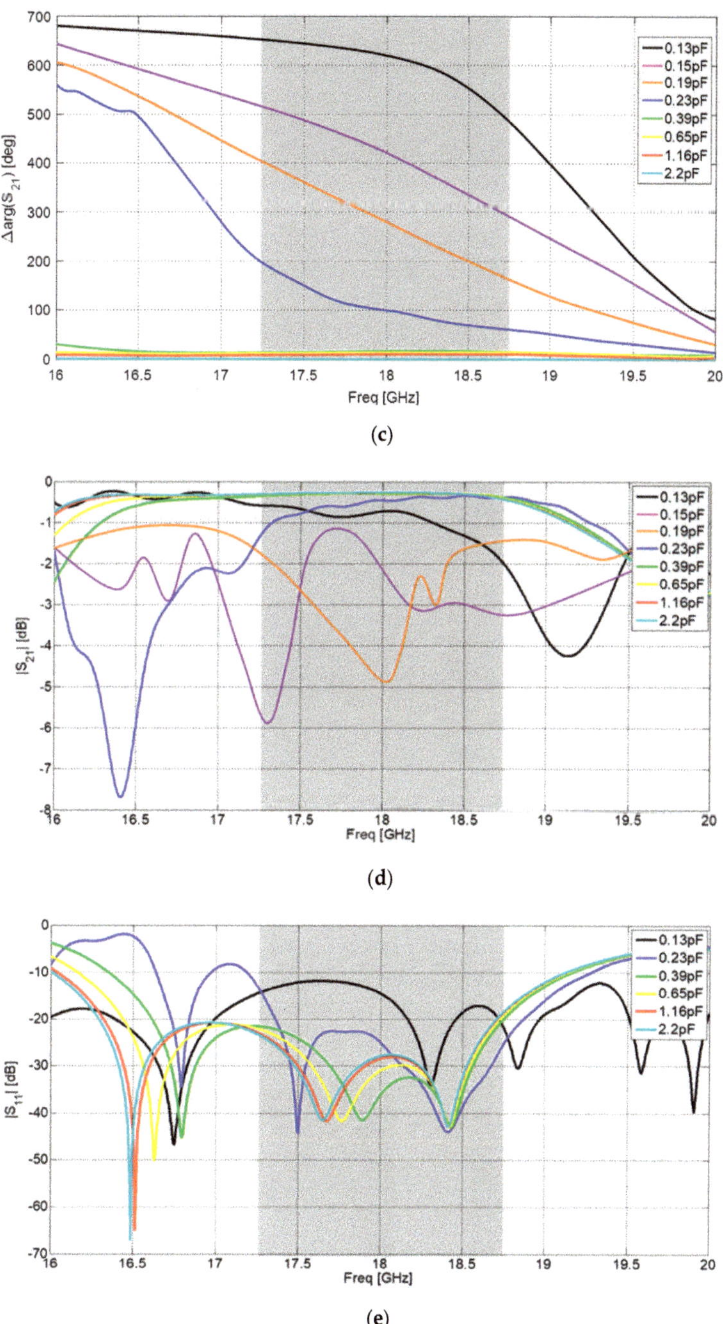

Figure 5. Simulation results of the 18 GHz phase shifter, (**a**) design; (**b**) arg(S_{21}); (**c**) Δarg(S_{21}) (related to min arg(S_{21}) value, 2.2 pF); (**d**) |S_{21}|; (**e**) |S_{11}|.

The shifter design presents proper matching in the desired band and a loss level of around 6 dB in the worst case. Considering that the working band is centered at 18 GHz, this loss level can be considered an appropriate value, if compared to commercially available devices. The phase shifting capability of the shifter is around 600°, almost twice the phase variation obtained for the reflective circuit, as expected.

The high operating frequency band (central frequency 18 GHz) is highly demanding and limiting in terms of design and manufacturing. In our work, best efforts are concentrated in having an easy-manufacturing device with reasonable performance results. Thus, the device is based on planar microstrip technology, which is almost in its frequency limit of operation. All the manufacturing and mounting has been carried out in our laboratory. The photolithography manufacturing process introduces a manufacturing size uncertainty of around 100 microns. The substrate is a Neltec N9217 (0.508 mm thickness, relative permittivity of 2.17, and <0.0008 loss tangent). The lumped elements (the MA46585 varactors) are welded by means of conductive epoxy. The device circuit design dimensions are provided in Table 1.

Table 1. Device circuit design dimensions (mm).

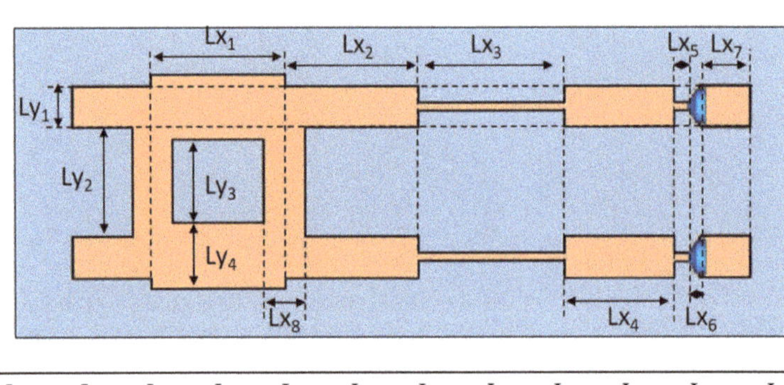

L_{x1}	L_{x2}	L_{x3}	L_{x4}	L_{x5}	L_{x6}	L_{x7}	L_{x8}	L_{y1}	L_{y2}	L_{y3}	L_{y4}
3.43	3.06	3.49	2.5	0.47	0.3	1	0.75	0.75	2.25	1.75	1.25

3. Results: Phase Shifter Prototype Performance

The design has been validated through a prototype. Figure 6.a yields some details of the shifter prototype, and the TRL (transmission-reflection-line) kit for the calibration prior to the measuring process. The measurement results are provided in the rest of the figures in Figure 6, in order to validate its proper functioning and performance.

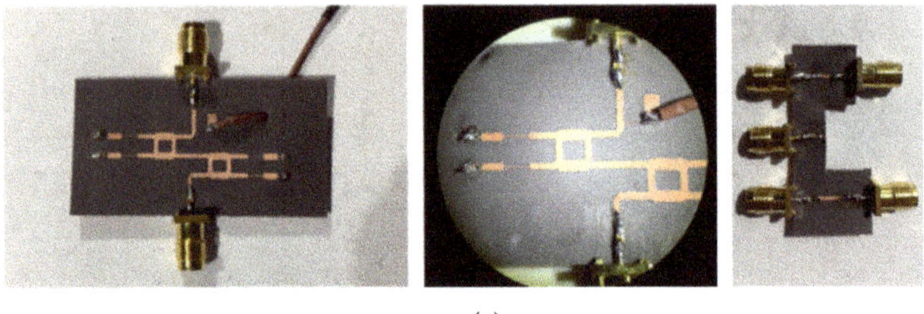

(a)

Figure 6. *Cont.*

Figure 6. *Cont.*

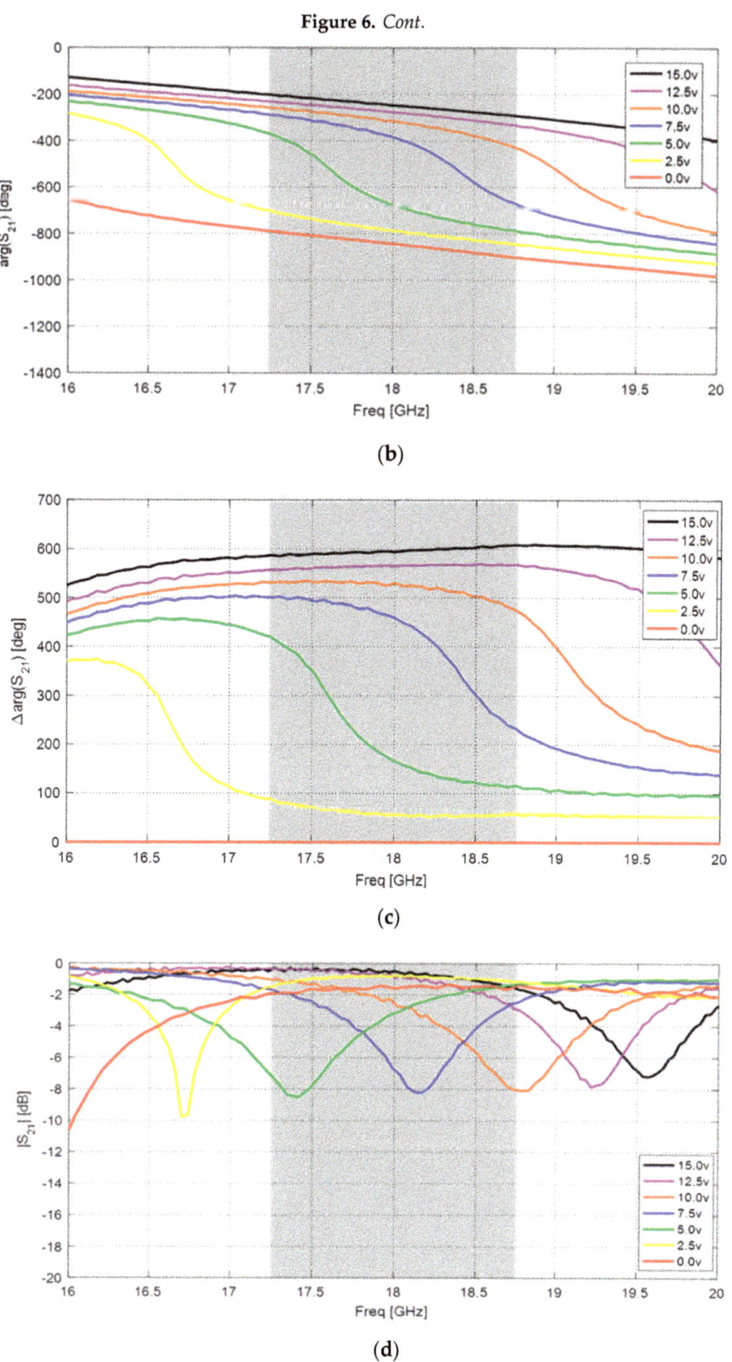

(**b**)

(**c**)

(**d**)

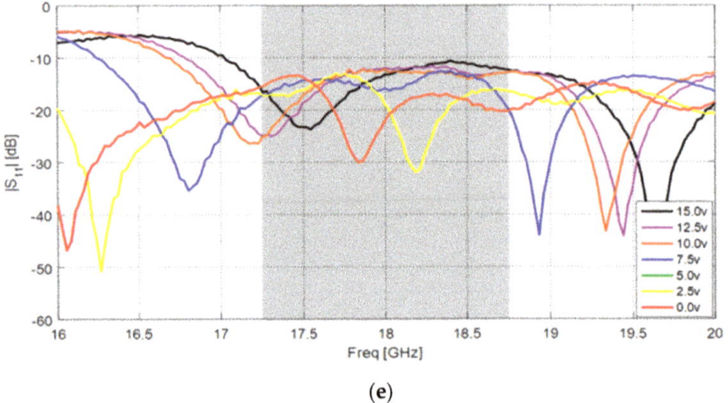

(e)

Figure 6. Prototype results of the 18 GHz phase shifter: (**a**) prototype and TRL kit (central image corresponds to a microscope zoom view); (**b**) arg(S_{21}); (**c**) Δarg(S_{21}) (related to min arg(S_{21}) value, 0 v); (**d**) |S_{21}|; (**e**) |S_{11}|.

As it can be noticed, there is a good agreement between the design performance and the prototype measurements. The insertion loss levels of the prototype are higher than the expected ones in the design. There exists a variety of possible causes for that: parasitic components of the varactor, conductive epoxy resistance, and roughness of the shifter printed line limits. For the sake of completeness, Figure 7 compares the simulation phase results with the experimental ones (Figure 7a), and provides an insight of the linearity of the device in terms of phase/voltage relation (Figure 7b).

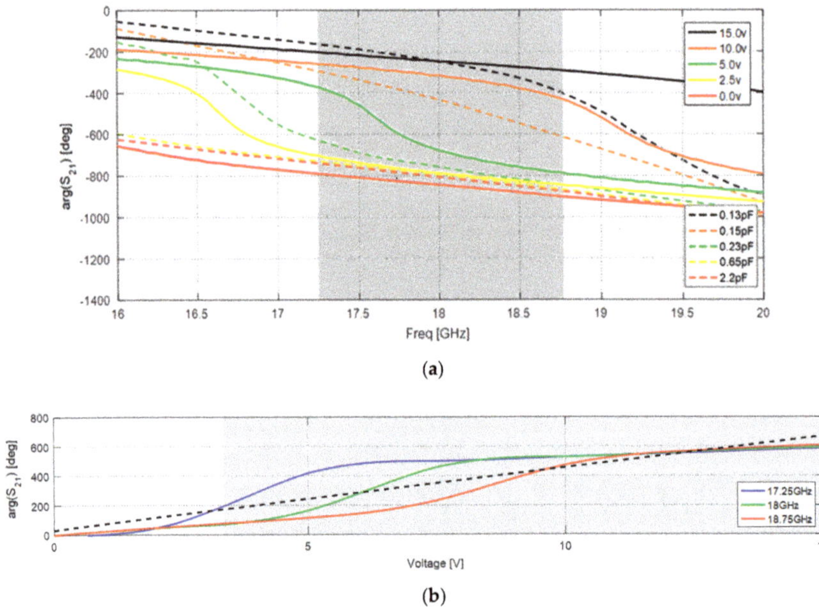

(a)

(b)

Figure 7. Phase shifting performance: (**a**) comparison between simulation and measurement phase results of the 18 GHz phase shifter; (**b**) phase/voltage relation (the dashed line is the linear reference and the grey zone is the one that provides at least 360° variation).

Notice that the simulation results in Figure 7a are in terms of capacitance and the experimental ones are in terms of voltage. It must be noticed that there is a non-negligible difference between pairs of simulation and experimental curves. However, the phase range of variation is preserved and the phase shifter prototype is fully functional, only needing a voltage/phase conversion table. In addition, it must be pointed out that, although the phase/capacitance relation is not linear, it is compensated by additional non-linear behavior of the capacitance/voltage relation. It is worth mentioning that, although the device provides a quasi linear phase/voltage response at the central frequency, this linearity is strongly degraded at the band limits, as it can be seen in Figure 7b. However, since the prototype provides a complete phase shift of around 600°, part of this possible variation can be sacrificed in order to increase the linearity of the device, as detailed in Figure 7b, where the grey region is the one that provides at least 360° phase variation.

For the sake of completeness, the next table (Table 2) provides a comparison of the performance of our device with the previously referenced literature works [10–15].

Table 2. Performance comparison of our device with the previously referenced literature works.

Work	Central Frequency	Bandwidth	Insertion Losses	Phase Shifting Range
[11]	2 GHz	200 MHz	<4.6 dB	240°
[12]	2 GHz	200 MHz	<1.6 dB	385°
[13]	2.5 GHz	500 MHz	<1.3 dB	150°
[14]	2.5 GHz	500 MHz	<1.2 dB	130°
[15]	12 GHz	>1 GHz	<3.2 dB	290°
[10]	12.5 GHz	>2 GHz	<3 dB	460°
This work	18 GHz	>1.5 GHz	<8 dB	600°

4. Conclusions

This communication presents the design and validation through prototyping of a microstrip phase shifter at 18 GHz for K band applications, with electronically controllable phase shift. The design is based on the use of loaded 3 dB/90° couplers. The performance of the reflective loads is improved due to the use of non-sequential impedance transformers. This design is particularly useful for reconfigurable phased arrays such as transmitarrays or reflectarrays, being fully integrable in such designs because of the reduced footprint. Furthermore, the shifter architecture is suitable for miniaturization, increasing the permittivity of the substrate. The dynamic phase range of the phase shifter is higher than 600° for the entire shifter bandwidth (>1.5 GHz). This phase range is higher than a complete phase turn of 360°, which allows the introduction of phase wrapping. The device losses are lower than 8 dB.

Author Contributions: M.T.E., A.P.-C., and A.A.-A. designed and simulated the phase shifter. P.E., C.S.-G., and P.P. manufactured and measured the prototype. A.P.-C. and P.P. wrote the document. J.F.V.-V. and P.P. supervised the whole study. All the authors participated in revising the article.

Funding: This work has been partially supported by the TIN2016-75097-P, RTI2018-102002-A-I00, and EQC2018-004988-P projects of the Spanish National Program of Research, Development, and Innovation and project B-TIC-402-UGR18 of Junta de Andalucía.

Conflicts of Interest: The authors declare no conflict of interest.

References

1. Bhattacharyya, A.K. *Phased Array Antennas: Floquet Analysis, Synthesis, BFNs and Active Array Systems*, 1st ed.; Wiley-Interscience: Newark, NJ, USA, 2006.
2. Bhartia, P.; Bahl, I.; Garg, R.; Ittipiboon, A. *Microstrip Antenna Design Handbook*; Artech House Publishers: Norwood, MA, USA, 2000.
3. Lee, K.; Tong, K. Microstrip Patch Antennas—Basic Characteristics and Some Recent Advances. *Proc. IEEE* **2012**, *100*, 2169–2180.

4. Carrasco, E.; Barba, M.; Encinar, J.A. Aperture-coupled reflectarray element with wide range of phase delay. *Electron. Lett.* **2006**, *42*, 667–668. [CrossRef]

5. Padilla, P.; Muñoz-Acevedo, A.; Sierra-Castañer, M.; Sierra-Pérez, M. Electronically reconfigurable transmitarray at Ku band for microwave applications. *IEEE Trans. Antennas Propag.* **2010**, *58*, 2571–2579. [CrossRef]

6. Hum, S.V.; Perruisseau-Carrier, J. Reconfigurable Reflectarrays and Array Lenses for Dynamic Antenna Beam Control: A Review. *IEEE Trans. Antennas Propag.* **2014**, *62*, 183–198. [CrossRef]

7. Lau, J.Y.; Hum, S.V. A Planar Reconfigurable Aperture With Lens and Reflectarray Modes of Operation. *IEEE Trans. Microw. Theory Tech.* **2010**, *58*, 3547–3555. [CrossRef]

8. Padilla, J.L.; Padilla, P.; Valenzuela-Valdés, J.F.; Fernández, J.M. High-frequency radiating element and modified 3dB/90° electronic shifting circuit with circular polarization for broadband reflectarray device cells. *Electron. Lett.* **2014**, *50*, 1042–1043. [CrossRef]

9. Silva, J.S.; Lima, E.B.; Costa, J.R.; Fernandes, C.A.; Mosig, J.R. Tx-Rx Lens-Based Satellite-on-the-Move Ka-Band Antenna. *IEEE Antennas Wirel. Propag. Lett.* **2015**, *14*, 1408–1411. [CrossRef]

10. Padilla, P.; Valenzuela-Valdés, J.F.; Padilla, J.L.; Fernández-González, J.M.; Sierra-Castañer, M. Electronically Reconfigurable Reflective Phase Shifter for Circularly Polarized Reflectarray Systems. *IEEE Microw. Wirel. Compon. Lett.* **2016**, *26*, 705–707. [CrossRef]

11. Lin, C.; Chang, S.; Chang, C.; Shu, Y. Design of a Reflection-Type Phase Shifter With Wide Relative Phase Shift and Constant Insertion Loss. *IEEE Trans. Microw. Theory Tech.* **2007**, *55*, 1862–1868. [CrossRef]

12. Burdin, F.; Iskandar, Z.; Podevin, F.; Ferrari, P. Design of Compact Reflection-Type Phase Shifters With High Figure-of-Merit. *IEEE Trans. Microw. Theory Tech.* **2015**, *63*, 1883–1893. [CrossRef]

13. An, B.; Chaudhary, G.; Jeong, Y. Wideband tunable phase shifter with low in-band phase deviation error using coupled line. *IEEE Microw. Wirel. Compon. Lett.* **2018**, *28*, 678–680. [CrossRef]

14. Chaudhary, G.; An, B.; Jeong, Y. In-band phase minimization method for wideband tunable phase shifter. *Microw. Opt. Technol. Lett.* **2019**, *61*, 537–541. [CrossRef]

15. ElKhorassani, M.T.; Vaquero, M.A.; Palomares, A.; Valenzuela-Valdés, J.F.; Padilla, P.; Touhami, N.A. Electronically tunable phase shifter with enhanced phase behaviour at Ku Band. In Proceedings of the 12th European Conference on Antennas and Propagation (EuCAP 2018), London, UK, 9–13 April 2018.

Article

Monitoring of Huntington's Disease Based on Wireless Sensing Technology

Qiyu Zhu [1], Lei Guan [2], Muhammad Bilal Khan [1] and Xiaodong Yang [1,*]

[1] School of Electronic Engineering, Xidian University, Xi'an 710071, China; qyzhu_xd@163.com (Q.Z.), engrmbkhan1986@gmail.com (M.B.K.)
[2] School of Life Science and Technology, Xidian University, Xi'an 710071, China; 15926395470@163.com
* Correspondence: xdyang@xidian.edu.cn

Received: 17 December 2019; Accepted: 22 January 2020; Published: 27 January 2020

Abstract: Huntington's disease (HD) is a rare genetic disorder that cannot be cured by current medical techniques. With the development of the disease, the life of patients will become more and more difficult. It is necessary to timely and effectively evaluate the development of the patient's condition based on the patient's clinical symptoms to help doctors to formulate a reasonable and effective treatment plan, alleviate the condition, and improve the quality of life, which reflects humane care. Currently, wearable devices or video surveillance are generally used to monitor the patients, and these schemes have some disadvantages. This paper presents a new method to monitor patients with HD using wireless sensing technology. Firstly, experimental data were collected by the self-developed microwave sensing platform (MSP), and then the data were preprocessed. Finally, support vector machine (SVM) and random forest (RF) algorithms were used to train the model. The MSP system continuously monitors patients in a non-contact way, which will not bring inconvenience to patients' lives, and will not involve privacy issues. The experimental results show that the prediction accuracy of SVM can be as high as 98.0% and that of RF can be as high as 96.7%, which proves the feasibility of the technical scheme described in this paper.

Keywords: HD; MSP; RF; SVM; wireless sensing technology

1. Introduction

Huntington's disease (HD) is a very rare autosomal dominant genetic disease. The cause of the disease is the mutation of Huntington gene on chromosome 4 of the patient, resulting in the variation of protein, which leads to the change of normal neural function through the related molecular mechanism [1]. The diagnosis of this disease depends on genetic testing. It usually develops around 35–45 years old [2]. The disease degenerates the patient's physical and psychological intelligence during the working-age and places a heavy burden on the patient. The clinical symptoms of HD mainly fall into three categories: motor symptoms, cognitive symptoms, and mental symptoms [3–5]. The typical manifestation of motor symptoms is that the fingers appear to play piano-like movements, accompanied by weird facial expressions. If the trunk is involved, the patient can have a dance-like gait. As the disease progresses, other body movements will also become slow and uncoordinated. In the end, the patient's entire system will be affected, making it difficult for the patient to complete simple daily actions such as walking, talking, eating, dressing, and washing. Cognitive symptoms sometimes appear many years earlier than motor symptoms. In the early stages of cognitive symptoms, patients do not only suffer from episodic memory, but also have significant functional dysfunction and do not understand the meaning of the speaker's language. As the disease progresses, the patient's dementia symptoms worsen. However, even if the condition is severe, the patient retains some cognitive functions. The third type of symptoms are mental symptoms. The mental symptoms of patients are often earlier than or synchronous with abnormal motor symptoms, mainly manifested as depression, often accompanied by insomnia, anorexia, and personality changes. In the later stage, patients will gradually experience hallucinations, delusions, paranoia, and aggressive behavior.

HD is characterized by complex clinical symptoms, progressive deterioration of the patient's condition, and usually death 15–20 years after onset [6]. At present, there is no effective treatment for this disease. The purpose of treatment is to improve the quality of life of patients. Several existing treatment methods can be classified as cause treatment and symptomatic treatment [7]. Cause treatment includes direct gene therapy and other indirect molecular therapy, this method cannot be realized at present, but a lot of research has been carried out. The symptomatic treatment is that the doctor obtains the clinical symptoms of the patient through visual observation and other methods and diagnoses the patient's condition against the HD Scale [8], so as to formulate a drug treatment plan to directly alleviate the patient's symptoms.

Drug therapy is currently the main treatment for HD. In order to confirm whether drug therapy is effective, doctors need to continue to follow up the effect of drug therapy, so as to improve the treatment. But this treatment has a disadvantage; the doctor's judgment of the patient is mainly based on the naked eye and the HD Scale. The diagnosis result can easily be affected by the subjective consciousness of the doctor. A promising solution is needed to continuously monitor the patient and provide an objective diagnosis basis. At the same time, the mental symptoms of patients are unstable, and patients are prone to hallucinations and aggressive behavior. In order to avoid self-injury, we also need to carry out continuous monitoring and give timely psychotherapy to patients.

In order to solve the above-mentioned problems, this paper proposes a new method to monitor HD patients using wireless sensing technology and demonstrates the feasibility of the proposed method. We independently developed the microwave sensing platform (MSP) monitoring tool, which is composed of a transmitter module and a receiver module. It can be directly installed in the indoor environment, without any contact with the patient, and it can be completely monitored in a non-contact way. The working process of MSP is: the transmitting module transmits the wireless signal in C-band, the receiving module receives the wireless signal and extracts the channel state information (CSI) data through channel estimation. After collecting CSI data, we first carried out a series of data preprocessing steps, including removing outliers and wavelet transform de-noising. Then we extracted the features of the preprocessed data to make the sample set. Finally, we used support vector machine (SVM) [9] and random forest (RF) [10] to train the classification model, respectively. The trained classification model can effectively distinguish the behavior of HD patients and normal people. The experimental results show that the accuracy of SVM and RF is 98.0% and 96.7%, respectively, which proves that the methods described in this paper can effectively monitor HD patients.

The rest of this paper is organized as follows. The Section 2 will introduce the current research on the monitoring of HD and make a brief comparison with the experimental scheme proposed in this paper. In the Section 3, we will introduce the principle of wireless sensing technology, and in the Section 4, we will describe the experimental scheme design. In the Section 5, we will describe the data processing flow. In the Section 6, we will discuss the experimental results and draw conclusions in the Section 7.

2. Related Work

At present, a large number of scholars have carried out related research on monitoring patients with HD using wearable devices. For example, Dinesh et al. reported a preliminary study to analyze motor symptoms associated with HD and Parkinson's disease based on sensor signal detection and data analysis. They use a light-weight, low-power sensor to monitor the motor symptoms of patients. The sensor adheres to the limbs and chest of patients like a tattoo and can continuously monitor patients for 48 hours. The experimental results show that the sensor can capture different clear signals of different clinical symptoms [11]. Bennasar et al. proposed an HD upper limb dyskinesia evaluation system. A triaxial acceleration sensor was worn on each wrist and chest of the experimental participants. The collected sensor data was used to develop an automatic classification system to distinguish normal people from HD patients [12,13] proposed a method of using inertial sensors to identify healthy gait, HD patient gait, and hemiplegic patient gait. This method is based on a supervised and trained two-state hidden Markov model, which can be extended to different research subjects for clinical practice and personal health assessment. At the same time, Francisco et al. collected the gait data of HD patients by binding the iPhone on the ankles of patients, using the built-in smart sensor of iPhone, and classified the data with the general assembly (meta) classifier algorithm to distinguish normal people and HD patients [14]. The use of wearable devices to monitor the movements and gaits of HD patients requires the binding of electronic devices to a part of the patient's body, which will affect the patient's movements to a certain extent and cannot collect movement data in the natural state of the patient. At the same time, it will also affect the patient's living comfort. Some scholars have also proposed the idea of using a camera to monitor patient activity, combined with related video image processing algorithms to extract patient activity information. For example, Agrawal et al. proposed a method for human fall detection based on video surveillance, but this may violate the privacy of patients [15].

As far as the authors know, this paper proposes for the first time to use C-band wireless sensing technology to monitor HD patients completely in a non-contact way for continuous monitoring. Compared with the traditional monitoring scheme, it has the following characteristics:

(1)   Monitoring patients completely in a non-contact way will not bring discomfort to the patient's body, and data can be collected under natural conditions.
(2)   Continuous monitoring of the patient's condition by wireless sensing will not affect the patient's privacy.
(3)   MSP has no requirements for the working environment and is easy to install.
(4)   The collected data are processed, and two machine learning algorithms are used to train the model. The two machine learning algorithms can be compared with each other, which makes the experimental results more convincing.

3. Principle of Wireless Sensing Technology

The proposed technical scheme is suitable for non-contact continuous monitoring of HD patients in an indoor environment. Therefore, this section will first describe the indoor model of wireless signal propagation in detail and explain the principles behind wireless sensing technology. Since this article mainly collects CSI data through MSP, we will next reveal the nature of CSI data, derive its mathematical expression, and describe MSP in detail at the end of this section.

3.1. Indoor Propagation Model of Wireless Signal

When the wireless signal propagates in the indoor environment, it is influenced by the obstacles in the propagation path, which cause reflection and diffraction. After the propagation of different paths, each component reaches the receiving end with different strengths and phases, resulting in a multipath effect [16]. The indoor propagation model of the wireless signal is shown in Figure 1.

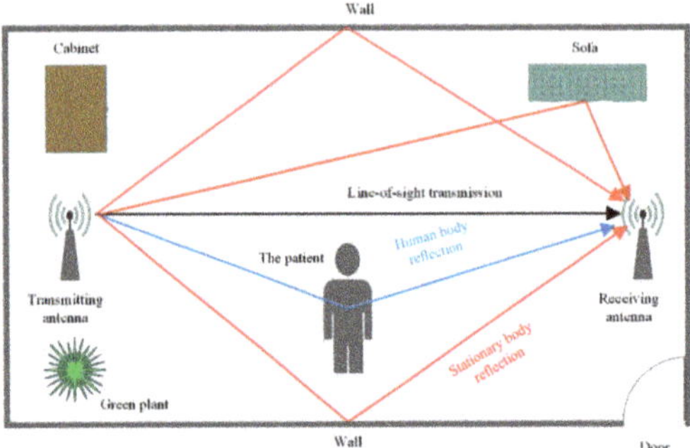

Figure 1. Indoor propagation model of wireless signal.

According to the Fries transfer formula [17], the receiving power of the receiving antenna can be expressed as:

$$P_r = \frac{P_t G_t G_r \lambda^2}{(4\pi R)^2} \tag{1}$$

where P_t and P_r are the power of transmitting antenna and receiving antenna, respectively; G_t and G_r are the gain of transmitting antenna and receiving antenna, respectively; and the distance between transmitting antenna and receiving antenna is R. λ is the wavelength of electromagnetic wave.

Assuming that the propagation path length of the electromagnetic wave through the interference of static objects is D, and the propagation path length through the interference of the human body is X, the Fries transfer formula can be rewritten as follows.

$$P_r = \frac{P_t G_t G_r \lambda^2}{16\pi^2 (R^2 + D^2 + X^2)} \tag{2}$$

In Formula (2), we can see that both R and D do not change. When people are indoors, X will change, resulting in a change in the power of the receiving antenna. At the same time, because the signal phase is a linear function of the distance of the propagation path, the change of the propagation path will also lead to the change of the signal phase [18]. Human behavior changes the strength and phase of the signal, and the CSI describes the loss and fading on the transmission path. When there is a moving target between the transmitting and receiving devices, the wireless signal reflected by the moving target increases the dynamic component of the channel. The fluctuation of the channel corresponds to the motion information of the target one by one. We can get CSI data from the signals collected by the receiving devices. By analyzing the CSI data, we can sense the changes of the external environment. In the next section, we explain the essence of CSI and its mathematical expression [19].

3.2. CSI

In order to eliminate the adverse effect of a multipath effect on wireless signal transmission, the MSP described in this paper uses orthogonal frequency division multiplexing (OFDM) modulation technology to decompose the data stream to be transmitted into several independent sub-data streams, that is, multiple subcarriers, and then transmits them in parallel, which can effectively eliminate the inter-symbol interference caused by the multipath effect in high-speed data stream transmission. At the same time, OFDM modulation technology can also greatly improve the data transmission efficiency. Because there are multiple subcarriers, each subcarrier channel is independently available, which also increases the amount of data to extract more information [20].

Multiple input multiple output (MIMO) is supported by OFDM modulation technology [21]. The channel model of the MIMO system in the frequency domain can be expressed as

$$Y = \mathbf{H}X + N \tag{3}$$

where Y represents the receiving signal, X represents the transmitting signal, N represents the environmental noise, and $\mathbf{H}$ represents the state matrix of the wireless channel, and its dimensions are $N_T \times N_R \times N_C$. N_T, N_R, and N_C, respectively, represent the number of transmitting antennas, receiving antennas, and subcarriers.

CSI is essentially a representation of the frequency response of each subcarrier channel. For each independent subcarrier channel, its frequency response can be expressed as

$$H_k(f_k) = \|H_k(f_k)\|e^{jarg(H_k(f_k))} \ (1 \le k \le N_C) \tag{4}$$

where f_k represents the center frequency of the Kth subcarrier, $\|H_k(f_k)\|$ represents the CSI amplitude information of the Kth subcarrier, and $arg(H_k(f_k))$ represents the CSI phase information of the Kth subcarrier.

After continuous data collection over a period of time, CSI data can be obtained through the channel estimation formula [22].

$$\hat{H} \approx \frac{Y}{X} \tag{5}$$

3.3. MSP

The MSP independently developed in this paper works in C-band (4.8 GHz). It is a highly customizable platform that can adapt to different application scenarios. MSP consists of omnidirectional antenna, industrial personal computer, absorbing material, frequency converter, and other related facilities. The use of absorbing materials is mainly to shield the surrounding environment from interference. The main function of MSP is to obtain the CSI of the wireless channel. By analyzing the CSI data, the behavior of patients can be monitored.

MSP uses OFDM technology. Its essence is an OFDM transceiver system, and its functional block diagram for obtaining CSI is shown in Figure 2.

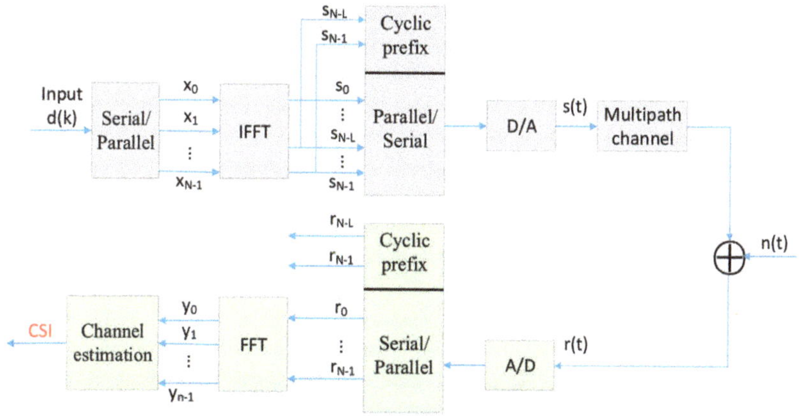

Figure 2. Block diagram of microwave sensing platform (MSP) to obtain channel state information (CSI) data.

In Figure 2, $d(k)$ is converted to N parallel data $\{x_0, x_1, \ldots, x_{N-1}\}$ through serial–parallel conversion. These data can be regarded as N data in the frequency domain. A set of time domain data

$\{s_0, s_1, \ldots, s_{N-1}\}$ obtained after the inverse discrete Fourier transform (IDFT) is an OFDM symbol. After adding a cyclic prefix to an OFDM symbol, the OFDM symbols are transmitted on the wireless multipath channel after parallel–serial conversion and digital–analog conversion. At the receiving end, the reverse work is performed: analog–digital conversion, parallel–serial conversion, removing cyclic prefix, and fast Fourier transform (FFT). The training sequence after FFT is used to perform channel estimation according to Equation (5), and CSI data can be obtained.

4. The Experimental Scheme

The purpose of this article is to continuously monitor patients with HD. To distinguish the patient's actions from other normal daily movements, so as to provide an objective clinical diagnosis basis for doctors and facilitate doctors to make appropriate treatment plans. The experimental flow chart is shown in Figure 3.

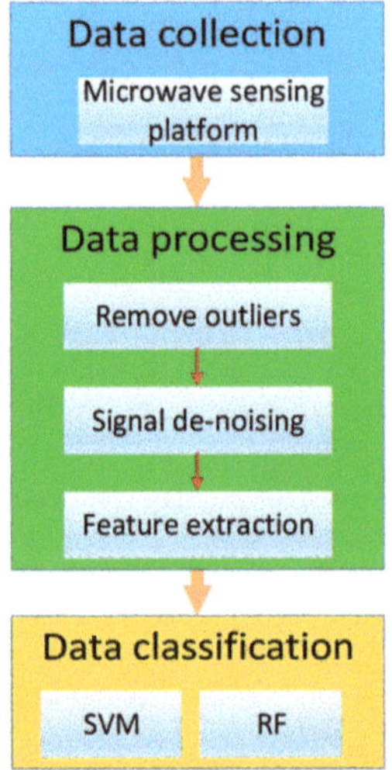

Figure 3. The experiment flow chart.

This experiment will collect the data of several actions like normal standing, normal sitting, normal walking, standing of HD, sitting of HD, and gait of HD, as shown in Table 1.

Table 1. The actions used in the experiment.

No.	Action
1	Normal standing
2	Normal sitting
3	Normal walking
4	Standing of HD
5	Sitting of HD
6	Gait of HD

The simulation experiment was carried out in a laboratory in the new science and technology building of Xidian University, which is 7 m × 5 m in size. The transmitting and receiving antennas of the MSP were respectively placed at two ends of the laboratory, with a horizontal distance of 4 m. The transmitting antenna and receiving antenna were positioned 1.8 m from the ground and 1.2 m from the ceiling. The transmitting module of the MSP was composed of a control computer with a wireless adapter and an omnidirectional antenna. The wireless adapter was configured in an injection mode for sending wireless signals. The receiving module of the MSP consisted of a control computer with a wireless adapter and three omnidirectional antennas. The wireless adapter was configured in a listening mode for receiving signals and extracting CSI data. The MSP operated in C-band (4.8 GHz) and used OFDM technology to modulate the signal with a total of 30 subcarriers. The signal bandwidth was set at 20 MHz. The transmitting antenna had a packet frequency of 200 Hz, and the time window was 12 s. We collected 300 samples for each action. The experimental scene is shown in Figure 4.

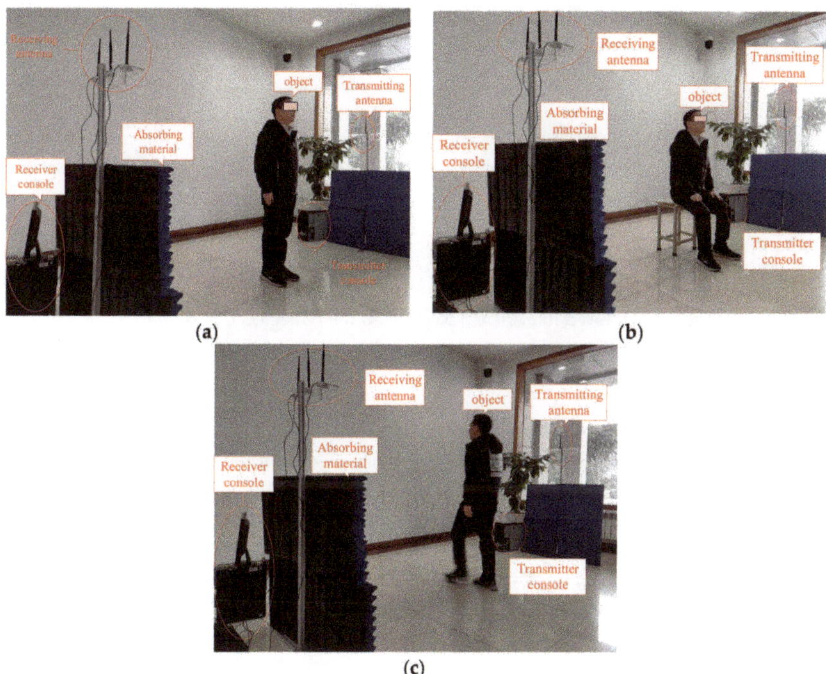

(a)

(b)

(c)

Figure 4. The experimental scene: (**a**) standing; (**b**) sitting; (**c**) walking.

In Figure 4c, the object moves in a direction perpendicular to the line connecting the transmitter and receiver, mainly considering that if the object moves along the line from the transmitter to the receiver, the line-of-sight transmission will be weakened, and most of the energy will be lost, resulting in a reduction of signal amplitude at the receiver, which is not conducive to any kind of communication. At the same time, in clinical trials, the movement tasks of patients are usually specified, so we chose a movement mode that is more conducive to study the feasibility of the technical scheme described in this article.

In this work, our research focuses on the feasibility of using wireless sensing technology to monitor HD patients, and the number of HD patients is very small [23]; hence, we did not recruit real patients, but volunteers from our team simulated HD movements.

Before the experiment, we fully informed the experimental participants of all relevant matters and contents of the clinical experiment. By watching videos of clinical manifestations of HD patients and reading related literature, all experimental participants were rigorously trained to simulate real HD patient movements.

Due to the limited staff in our team, there were a total of 10 volunteers who participated in our experiment, including 6 males and 4 females, aged between 24 and 48 years. Details of volunteers are shown in Table 2.

Table 2. Details of volunteers.

ID	Age	Gender	Weight (kg)	Height (cm)
1	26	Male	72	181
2	45	Male	53	173
3	30	Male	64	168
4	32	Female	50	155
5	24	Female	56	165
6	25	Male	62	164
7	36	Male	70	174
8	48	Male	65	175
9	25	Female	51	161
10	40	Female	57	163

We randomly selected 5 volunteers (3 male and 2 female) to simulate the movements of HD patients, and the remaining 5 volunteers (3 male and 2 female) were used as normal reference. Each set of actions was repeated 60 times for each volunteer, and it took 6 days to collect the data.

5. The Data Processing

After collecting CSI data, we needed to perform a series of processing steps on the data. In this section, we follow the steps shown in Figure 3 to process the collected CSI data in turn.

5.1. Data Preprocessing

5.1.1. Remove Outliers

In the process of data collection, due to the influence of environmental noise or internal voltage fluctuation of the device, there was a large number of outliers in the original signal. These outliers seriously distort the original signal and must be removed.

We used the "Hampel" function in MATLAB to remove the outliers of the original signal. For each sampling point of the original signal, the function calculates the median of the window consisting of the sampling point and the three sampling points on the left and right sides. Then the absolute value of the median is used to estimate the standard deviation of the median at each sampling point. If a sample is more than three standard deviations away from the median, the sample is replaced with the median [24]. The outliers contained in the original signal are shown in Figure 5. The original signal after removing the outliers is shown in Figure 6.

Outliers can affect signal de-noising. After removing the outliers of the original signal, we can de-noise the signal, which is described in the next section.

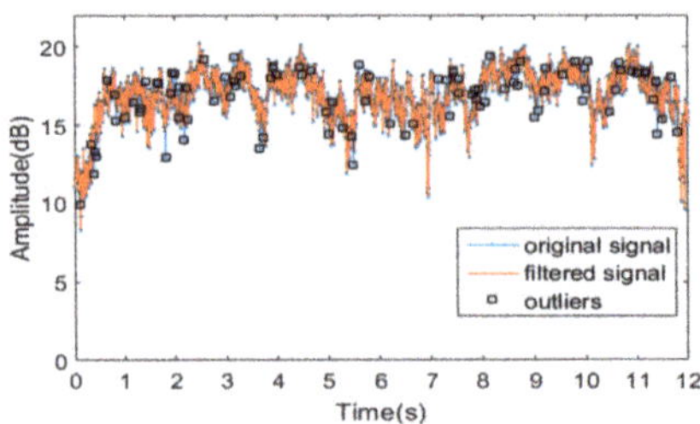

Figure 5. Outliers contained in the original signal.

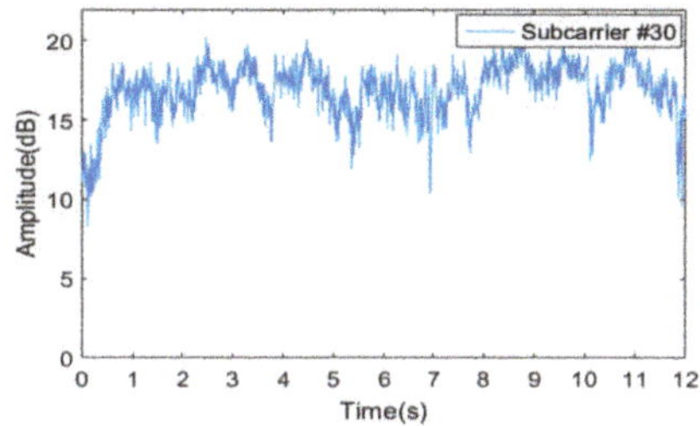

Figure 6. The original signal after removing the outliers.

5.1.2. Signal De-Noising

The patient's actions mainly affect the low-frequency components of the wireless signal, so we needed to filter out the high-frequency noise generated by environmental noise, slight internal voltage fluctuations, etc. In this paper, wavelet transform was used to realize signal de-noising [25]. The "wden" function with one-dimensional noise reduction in the MATLAB toolbox was used. The main principle of this function is to filter out noise through threshold processing of the wavelet decomposition coefficient of the original signal.

We used the "sym8" wavelet to decompose the original signal in 5 layers. The "SimN" (N = 2, 3, ... , 8) wavelet has good symmetry, which can reduce the phase distortion during signal decomposition and reconstruction to a certain extent. At the same time, we applied heuristics to overcome the problem of noise distribution at each decomposition level. The signal waveforms of each action after de-noising by wavelet transform are shown in Figure 7; the larger the variance, the larger the information. We chose the subcarrier according to the principle of maximum variance [26].

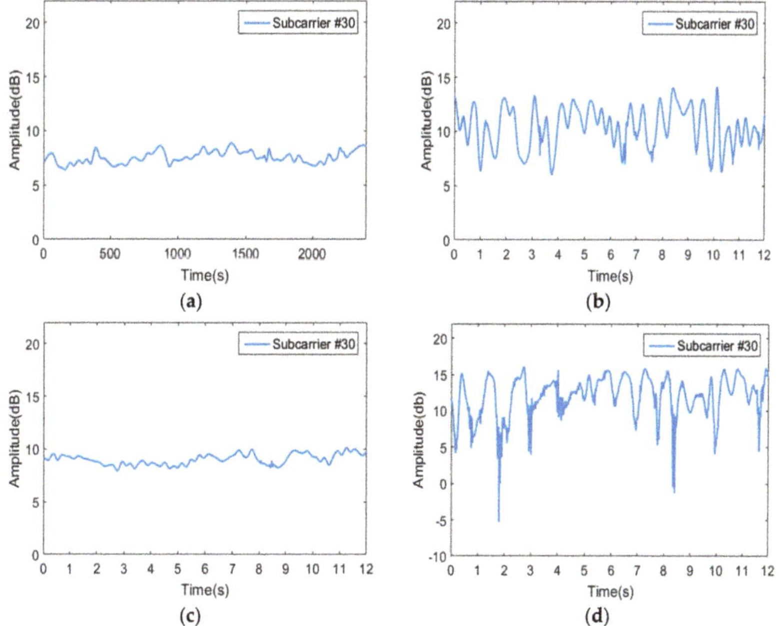

Figure 7. *Cont.*

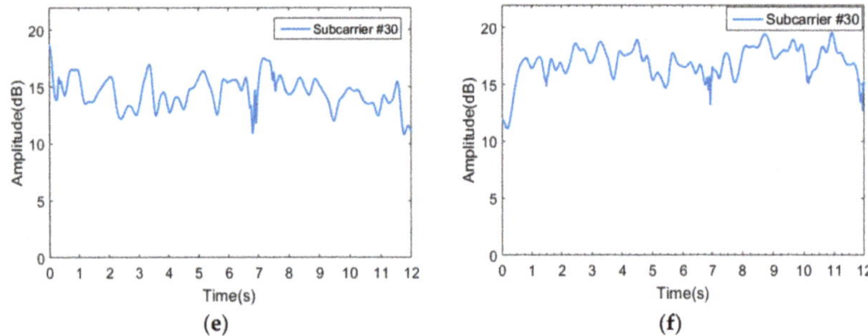

Figure 7. The waveforms of each action after data preprocessing: (**a**) normal standing; (**b**) standing of Huntington's disease (HD); (**c**) normal sitting; (**d**) sitting of HD; (**e**) normal walking; (**f**) gait of HD.

5.2. Feature Extraction

It can be seen from Figure 7 that the time domain waveform of each action is quite different. We extracted eight time domain features from the signal waveform of each action, as shown in Table 3.

Table 3. Time domain features.

Features	A Formula to Calculate
Mean value	$Y_{MV} = \frac{1}{N}\sum\limits_{i=1}^{N} x_i$
Standard deviation	$Y_{SD} = \sqrt[2]{\frac{1}{N-1}\sum\limits_{i=1}^{N}(x_i - Y_{MV})^2}$
Root mean square	$Y_{RMS} = \sqrt[2]{\frac{1}{N}\sum\limits_{i=1}^{N} x_i^2}$
Peak-to-peak value	$Y_{PPV} = \max(x_i) - \min(x_i)(i = 1,2,\ldots,N)$
Kurtosis	$Y_K = \dfrac{\frac{1}{N}\sum_{i=1}^{N}\left(\left\| x_i \right\| - Y_{MV}\right)^4}{Y_{RMS}^4}$
Skewness	$Y_S = \dfrac{\frac{1}{N}\sum_{i=1}^{N}\left(\left\| x_i \right\| - Y_{MV}\right)^3}{Y_{RMS}^3}$
Peak factor	$Y_P = \dfrac{\max(x_i)}{Y_{RMS}}(i = 1,2,\ldots,N)$
Waveform factor	$Y_W = \dfrac{N*Y_{RMS}}{\sum_{i=1}^{N}\left\| x_i \right\|}(i = 1,2,\ldots,N)$

5.3. Model Training

We used SVM and RF to train the model to ensure the accuracy of data classification and to determine which algorithm has a better effect in practical applications. At the same time, in order to make the training model reliable, we used the four-fold cross validation method to divide the data set.

We selected the radial basis function (RBF) as the kernel function of SVM, and the RF contained 500 decision trees.

6. Result and Discussion

The confusion matrix of each classification algorithm is shown in Table 4.

Table 4. Confusion matrix of classification algorithms.

Classification Algorithm	Actual Action	Predict Action (Number of Samples)					
		Standing	Sitting	Walking	Standing of HD	Sitting of HD	Gait of HD
SVM	Standing	73	0	2	0	0	0
	Sitting	1	74	0	0	0	0
	Walking	2	0	72	0	0	1
	Standing of HD	0	0	0	75	0	0
	Sitting of HD	0	0	0	0	75	0
	Gait of HD	0	0	1	0	2	72
RF	Standing	72	1	0	2	0	0
	Sitting	0	75	0	0	0	0
	Walking	0	0	75	0	0	0
	Standing of HD	4	0	0	71	0	0
	Sitting of HD	0	0	0	1	70	4
	Gait of HD	0	0	2	0	1	72

The experimental accuracy of each algorithm is shown in Figure 8.

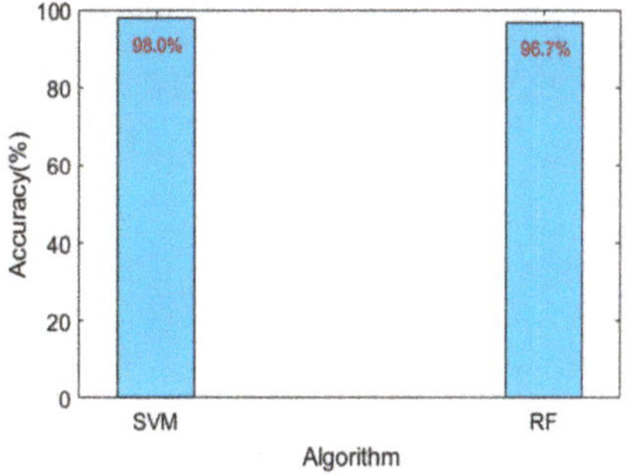

Figure 8. Algorithm accuracy.

Figure 7a–d shows different waveforms of normal people and patients with HD under static action. It can be seen from the figure that the static action signal waveform of normal people is relatively gentle, while the static action signal waveform of patients fluctuates greatly because patients with HD have convulsions when they are ill and dance-like movements when they are serious. Figure 7e,f shows the signal waveforms of normal gait and gait in patients with HD. Due to the large amplitude of walking itself, the abnormal body swing of patients with HD may be covered by the walking movement, resulting in the signal waveform discrimination between normal gait and abnormal gait not being very obvious, which is consistent with reality. At the same time, we can see from Figure 7 that the differences between the action signal waveform of normal people and patients with HD are obvious in the time domain. In order to reduce the amount of data and improve the efficiency of the algorithm training model, we extracted eight time domain features from the samples, which can well describe the time domain waveform, and the experimental results also prove this.

Table 4 is the confusion matrix of SVM and RF. It can be seen that the SVM algorithm can completely distinguish the static actions of patients with HD, while the normal static actions have misclassification, which shows that the performance of patients in the static actions is inconsistent, the performance of body convulsion or dance is different, and the performance of normal people in static actions are consistent. It can also be seen that RF can completely distinguish between normal gait and normal sitting action. Both SVM and RF cannot distinguish the normal gait and the abnormal gait of patients with HD completely, because the time domain signal waveform of them is similar.

Figure 8 shows the prediction accuracy of the two algorithms. The prediction accuracy of SVM is 98.0%, and that of RF is 96.7%. Both algorithms can achieve high prediction accuracy, which proves that the experimental scheme described in this paper is feasible. At the same time, we can draw a conclusion that when using the method described in this paper to distinguish the actions of patients with HD, the performance of SVM is better than that of RF.

7. Conclusions

As far as we know, this article first proposed a method for monitoring HD patients using wireless sensing technology and studied the feasibility of the proposed technical scheme in depth. We used the self-developed MSP to collect CSI data, then removed the outliers, and filtered the CSI data with the wavelet transform. After that, we extracted eight time domain features from each action data set and trained the model with SVM and RF machine learning algorithms. The experimental results show that the prediction accuracy of the SVM algorithm can reach 98.0%, and the prediction accuracy of the RF algorithm can reach 96.7%. Both algorithms can effectively distinguish the normal actions and the actions of patients with HD, which proves that the technical scheme described in this paper is feasible. This will provide a basis for doctors to objectively diagnose patients' conditions. At the same time, the technology can help doctors to follow up on the patient's condition development and improve the treatment plan in time. At present, the MSP described in this paper is not fully automated but also needs some manual operations. Next, we will continue to improve the experimental platform to make it fully automated to achieve data collection, data processing, and data analysis in a one-click operation. At the same time, we will also further explore the application of wireless sensing technology in medical care.

Author Contributions: Conceptualization, Q.Z.; methodology, L.G.; resources, X.Y.; writing—original draft preparation, Q.Z.; writing—review and editing, M.B.K.; supervision, X.Y.; project administration, X.Y.; funding acquisition, X.Y. All authors have read and agreed to the published version of the manuscript.

Funding: This research was funded by the National Natural Science Foundation of China grant number 61301175.

Conflicts of Interest: The authors declare no conflict of interest.

References

1. Oliveira, J.M.A. Nature and cause of mitochondrial dysfunction in Huntington's disease: focusing on huntingtin and the striatum. *J. Neurochem.* **2010**, *114*, 1–12. [CrossRef] [PubMed]
2. Walker, F.O. Huntington's disease. *Semin. Neurol.* **2007**, *27*, 143–150. [CrossRef] [PubMed]
3. Shannon, K.M. Huntington's disease—Clinical signs, symptoms, presymptomatic diagnosis, and diagnosis. *Handb. Clin. Neurol.* **2011**, *100*, 3–13. [PubMed]
4. Goodman, L.V.; Cha, J.; Como, P. Poster 24: Expert Treatment Preferences for the Motor, Mood, and Behavioral Symptoms of Huntington's Disease. *Neurotherapeutics* **2009**, *6*, 211–212.
5. Hofmann, N. Understanding the Neuropsychiatric Symptoms of Huntington's Disease. *J. Neurosci. Nurs.* **1999**, *31*, 309–313. [CrossRef] [PubMed]
6. Brackenridge, C.J. The relation of type of initial symptoms and line of transmission to ages at onset and death in Huntington's disease. *Clin. Genet.* **1971**, *2*, 287–297. [CrossRef] [PubMed]
7. Bonelli, R.M.; Hofmann, P. A systematic review of the treatment studies in Huntington's disease since 1990. *Expert Opin. Pharmacother.* **2007**, *8*, 141–153. [CrossRef] [PubMed]
8. Huntington Study Group. Unified Huntington's disease rating scale: Reliability and consistency. *Mov. Disord.* **1996**, *11*, 136–142. [CrossRef] [PubMed]
9. Widodo, A.; Yang, B.-S. Support vector machine in machine condition monitoring and fault diagnosis. *Mech. Syst. Signal Process.* **2008**, *21*, 2560–2574. [CrossRef]
10. Breiman, L. Random Forests. *Mach. Learn.* **2001**, *45*, 5–32. [CrossRef]
11. Dinesh, K.; Xiong, M.; Adams, J.; Dorsey, R.; Sharma, G. Signal analysis for detecting motor symptoms in Parkinson's and Huntington's disease using multiple body-affixed sensors: A pilot study. In Proceedings of the 2016 IEEE Western New York Image and Signal Processing Workshop (WNYISPW), Rochester, NY, USA, 18 November 2016; pp. 1–5.

12. Bennasar, M.; Hicks, Y.A.; Clinch, S.P.; Jones, P.; Holt, C.; Rosser, A.; Busse, M. Automated Assessment of Movement Impairment in Huntington's Disease. *IEEE Trans. Neural Syst. Rehabil. Eng.* **2018**, *26*, 2062–2069. [CrossRef] [PubMed]
13. Mannini, A.; Trojaniello, D.; della Croce, U.; Sabatini, A.M. Hidden Markov model-based strategy for gait segmentation using inertial sensors: Application to elderly, hemiparetic patients and Huntington's disease patients. In Proceedings of the 2015 37th Annual International Conference of the IEEE Engineering in Medicine and Biology Society (EMBC), Milan, Italy, 25–29 August 2015; pp. 5179–5182.
14. Acosta-Escalante, F.D.; Beltrán-Naturi, E.; Boll, M.C.; Hernández Nolasco, J.A.; García, P.P. Meta-Classifiers in Huntington's Disease Patients Classification, Using iPhone's Movement Sensors Placed at the Ankles. *IEEE Access* **2018**, *6*, 30942–30957. [CrossRef]
15. Agrawal, S.C.; Tripathi, R.K.; Jalal, A.S. Human-fall detection from an indoor video surveillance. In Proceedings of the 2017 8th International Conference on Computing, Communication and Networking Technologies (ICCCNT), Delhi, India, 3–5 July 2017; pp. 1–5.
16. Ali, A.H.; Razak, M.R.A.; Hidayab, M.; Azman, S.A.; Jasmin, M.Z.M.; Zainol, M.A. Investigation of indoor WIFI radio signal propagation. In Proceedings of the 2010 IEEE Symposium on Industrial Electronics and Applications (ISIEA), Penang, Malaysia, 3–5 October 2010; pp. 117–119.
17. Wang, C.; Chen, S.; Yang, Y.; Hu, F.; Liu, F.; Wu, J. Literature review on wireless sensing-Wi-Fi signal-based recognition of human activities. *Tsinghua Sci. Technol.* **2018**, *23*, 203–222. [CrossRef]
18. Adib, F.; Hsu, C.-Y.; Mao, H.; Katabi, D.; Durand, F. Capturing the human figure through a wall. *ACM Trans. Graph.* **2015**, *34*, 1–13. [CrossRef]
19. Xi, W.; Huang, D.; Zhao, K.; Yan, Y.; Cai, Y.; Ma, R.; Chen, D. Device-Free Human Activity Recognition Using CSI. In Proceedings of the 1st Workshop ACM, Denver, CO, USA, 16 October 2015.
20. Siriwongpairat, W.P.; Su, W.; Olfat, M.; Liu, K.J.R. Multiband-OFDM MIMO coding framework for UWB communication systems. *IEEE Trans. Signal Process.* **2006**, *54*, 214–224. [CrossRef]
21. Khan, M.B.; Yang, X.; Ren, A.; Al-Hababi, M.A.M.; Zhao, N.; Guan, L.; Fan, D.; Shah, S.A. Design of Software Defined Radios Based Platform for Activity Recognition. *IEEE Access* **2019**, *7*, 31083–31088. [CrossRef]
22. Edfors, O.; Sandell, M.; van de Beek, J.J.; Wilson, S.K.; Borjesson, P.O. OFDM channel estimation by singular value decomposition. *IEEE Trans. Commun.* **1998**, *46*, 931–939. [CrossRef]
23. Leung, C.M.; Chan, Y.W.; Chang, C.M.; Yu, Y.L.; Chen, C.N. Huntington's disease in Chinese: a hypothesis of its origin. *J. Neurol. Neurosurg. Psychiatry* **1992**, *55*, 681–684. [CrossRef] [PubMed]
24. Zhang, L.; Qin, Y.; Zhang, J. Study of polynomial curve fitting algorithm for outlier elimination. In Proceedings of the 2011 International Conference on Computer Science and Service System (CSSS), Nanjing, China, 27–29 June 2011; pp. 760–762.
25. Zhang, J.X.; Zhong, Q.-H.; Dai, Y.-P.; Liu, Z. A new de-noising method based on wavelet transform and transforming Hampel filter. In Proceedings of the SICE Conference IEEE, Fukui, Japan, 4–6 August 2003.
26. Yuan, L.; Kesavan, H.K. Minimum entropy and information measure. *IEEE Trans. Syst. Man Cybern. Part C Appl. Rev.* **1998**, *28*, 488–491. [CrossRef]

Article

Design of a Multi-Band Microstrip Textile Patch Antenna for LTE and 5G Services with the CRO-SL Ensemble

Carlos Camacho-Gomez, Rocio Sanchez-Montero *, Diego Martínez-Villanueva, Pablo-Luís López-Espí and Sancho Salcedo-Sanz

Department of Signal Processing and Communications, Escuela Politécnica Superior, Universidad de Alcalá, Campus Universitario, Ctra. de Madrid a Barcelona km 33.600, 28805 Alcalá de Henares, Spain; carlos.camacho@uah.es (C.C.-G.); diego.martinezv@edu.uah.es (D.M.-V.); pablo.lopez@uah.es (P.-L.L.-E.); sancho.salcedo@uah.es (S.S.-S.)
* Correspondence: rocio.sanchez@uah.es; Tel.: +34-91-8856660

Received: 17 January 2020; Accepted: 7 February 2020; Published: 9 February 2020

Featured Application: A novel textile U-shaped with concentric annular slot antenna prototype for LTE and 5G services has been described. In the ground plane, a meander slot has been introduced to reduce the antenna dimensions. A new multi-method metaheuristic algorithm, the Coral Reefs Optimization with Substrate Layer CRO-SL, has been used to optimize the antenna parameters and improve its performance in the frequency bands of interest.

Abstract: A textile multi-band antenna for LTE and 5G communication services, composed by a rectangular microstrip patch, two concentric annular slots and a U-Shaped slot, is considered in this paper. In the ground plane, three sleeved meanders have been introduced to modify the surface current distribution, leading to a bandwidth improvement. The U-Shaped slot, the dual circular slots, and the meanders shape have been optimized by means of the Coral Reefs Optimization with Substrate Layer algorithm (CRO-SL). This population-based meta-heuristic approach is a kind of ensemble algorithm for optimization (multi-method), in which different search operators are considered within the algorithm. We show that the CRO-SL is able to obtain a robust multi-band textile antenna, including LTE and 5G frequency bands. For the optimization process, the CRO-SL is guided by means of a fitness function obtained after the antenna simulation by a specific simulation software for electromagnetic analysis in the high frequency range.

Keywords: antenna design; constrained optimization problems; coral reefs optimization algorithm; meta-heuristics

---

1. Introduction

In the last decade, a large variety of wireless enabled portable devices such as smartphones, tablets or laptops have been introduced. The implementation of new mobile technologies further increases the bandwidth requirements of wireless systems in order to cover recently allocated LTE and 5G frequency bands [1]. Recent research works have allowed us to develop the design of antennas using textile materials in the substrate, leading to devices called "wearable antennas" [2]. One of the main advantages of the antennas based on textile materials is that they can be manufactured using smart fabric and interactive textile systems [3], in which unobtrusive integration of electronic components increases functionality of the garment [4,5].

Recently, the implementation of different kind of antennas in wearables has been massive [6–9]. Microstrip patch antennas are frequently used in textile materials because of their many advantages, such as low profile, light weight, and conformity. However, these kind of antennas suffer from important issues in their design process (precise value and model of fabric dielectric constant for simulations, difficulty to glue metallic parts to textile materials, bending and moisture influence in antenna performance, etc.), causing severe limitations in their practical applications. This fact is even more dramatic in the frequency bands of modern communication systems based on LTE and 5G technologies. The number of parameters to be tuned in order to make the antenna feasible for working in LTE and 5G applications is usually very high. In these cases, classical optimization methods are no longer suitable, and the employment of advanced optimization algorithms (mainly meta-heuristic approaches, among others) has been shown to be very useful for antenna design [10–14].

In this paper, we propose a new model of microstrip textile patch antenna, a multi-band device that can be tuned for LTE and 5G services, among others. Specifically, the proposed antenna is composed by a rectangular microstrip patch with two concentric annular slots and a U-Shaped slot, with sleeved meanders introduced in the ground plane to modify the surface current distribution, leading to a bandwidth improvement [15–17]. This original shape allows the antenna to work accurately in several frequency bands, including LTE and 5G communication services. On the other hand, this specific shape also leads to a hard optimization problem, with a high number of real variables and constraints to be taken into account. Moreover, the proposed antenna has been simulated considering a wearable substrate, which makes the direct designing process even more difficult.

A meta-heuristic algorithm for optimization is then considered in order to obtain a good design of the antenna, with excellent properties of bandwidth in all the considered frequency bands. Therefore, we propose to use a version of the Coral Reefs Optimization (CRO) algorithm [18]—in this case, the version with the substrate layer (CRO-SL) approach [19]. The CRO is an evolutionary type algorithm which simulates all the processes occurring in a real coral reef in order to carry out the optimization of a given system (the textile antenna considered in this work in this case). The CRO-SL version has been successfully applied to a number of optimization problems [19], and it is able to combine different search patterns or strategies within a single population of potential solutions. In this case, we will show how this optimization scheme is able to obtain excellent results in the optimization process of the proposed antenna, tuning it for its use in LTE and 5G communication systems. In the experimental section of the paper, we detail the antenna design process and its simulation with specific software in order to evaluate the potential of the CRO-SL in this design problem.

The rest of the paper is structured in the following way: the next section presents in detail the proposed antenna design, characteristics, and variables to be optimized. Section 3 describes the CRO-SL algorithm used to optimize the textile antenna for LTE and 5G systems. Section 4 shows the experiments carried out to optimize the antenna, and the results obtained in simulation of the optimized device. Finally, Section 5 closes the paper with some final remarks on the research carried out in this paper.

2. Antenna Model

The proposed microstrip patch antenna is shown in Figure 1. It combines several key features to provide the desired operational performance at different frequencies, including LTE and 5G bands. Specifically, the proposed design comprises a rectangular microstrip patch with two concentric annular slots. Each ring has four spikes within it. This type of patch antenna is inspired by different works previously presented in the literature [20–24]. Additionally, the design contains a U-shaped aperture of rectangular slots to obtain the resonant frequency for the LTE band. U-slot patch antennas are well known mainly for their wide-band characteristics and are capable of providing other advantages, including dual band and triple-band operations, due to their ability to be implemented with other patch antenna shapes, such as circular, triangular, or rectangular shapes [25]. The proposed antenna is completed with three meander slots at the bottom of the ground plane, according to the study in reference [26]. Table 1 shows the variables involved in antenna design and the variable ranges considered in this case.

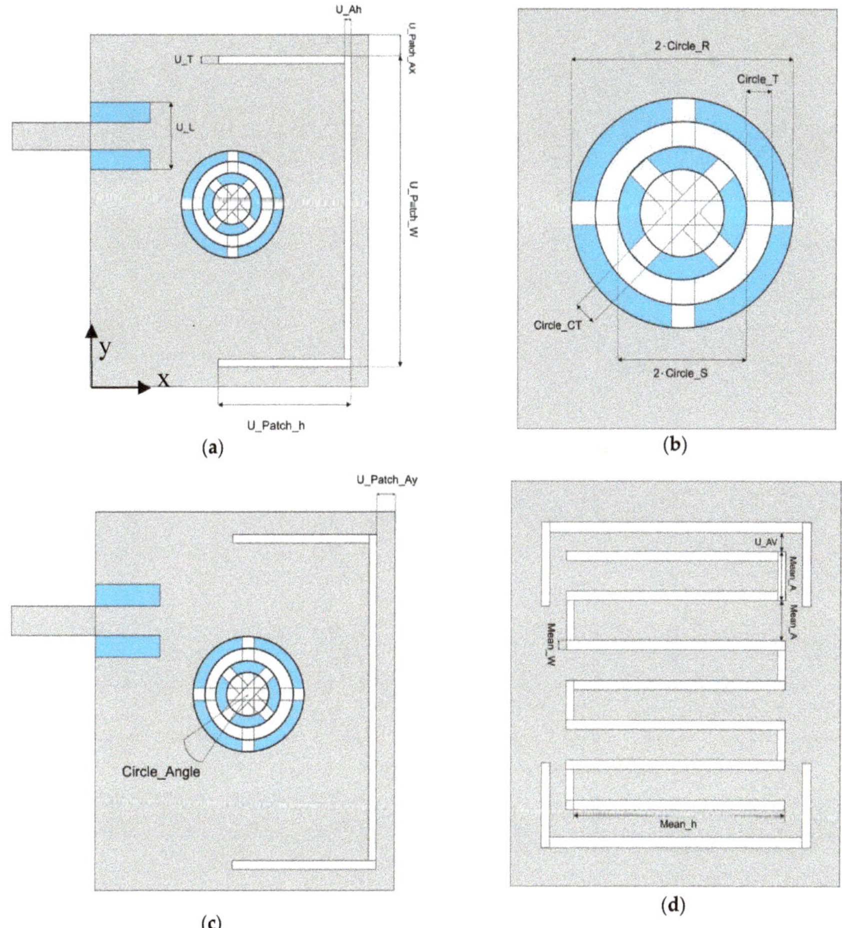

Figure 1. Variables description of the proposed antenna. (**a**) Antenna top view U-shaped variable description; (**b**) antenna top view the annular patch variable description; (**c**) detail description of the variables in the top view of the antenna that have been used in the optimization process; (**d**) antenna ground plane.

Note that the optimization of the proposed antenna in order to be operational in the desired frequencies is a hard problem in which metaheuristic algorithms can obtain good results. Note also that the objective function must be calculated starting from the antenna simulation for a given set of antenna parameters. Thus, the hybridization of a simulation software with a meta-heuristic algorithm is necessary to tackle the optimization of this antenna. Specifically, the CRO-SL algorithm will be used as optimizer, since its characteristics of multi-method with different search operators may work fine in this problem. The CRO-SL will be hybridized with the CST simulation software in order to obtain the performance of each antenna in the CRO-SL evolution.

Table 1. Definition of the variables to optimize in the design of the proposed antenna (see Figure 1 for details).

Variable	Range
Antenna top view	
U_T	[1,5]
U_Patch_h	[1,Lp]
U_Patch_AY	[0.5,(Wp-2Circle_R)/2]
U_Patch_AX	[0.5,Lp/2-U_Patch_W-Circle_R]
U_Ah	[1,(Wp-7Mean_A-8Mean_W-2U_T)/2]
U_L	[0.5,Wp]
U_Patch_W	[1,5]
Circle_R	[11,Wp/2]
Circle_Angle	[0.5,180]
Circle_S	[3,5]
Cirlce_CT	[1,5]
Circle_T	[3,5]
Lp=60mm; Wp=90mm	
Antenna ground plane	
Mean_W	[1,2.7]
Mean_A	[1,10]
Mean_h	[1,Lp]
U_AV	[1,(Lp-Mean_h-2U_T)/2]

3. Antenna Optimization: the CRO-SL Algorithm

A multi-method ensemble CRO-SL is the algorithm used to optimized the proposed antenna. This algorithm is an advanced version of a basic original version of the CRO [18]. The CRO is an evolutionary-type algorithm in which the search operators are based on the processes occurring in a coral reefs, including reproduction, fight for space, or depredation [19]. The pseudocode of the original CRO is shown below, with the different CRO phases (reef initialization and reef formation), along with all the operators applied to guide the search.

Table 2. Description of the pseudo-code for the Coral Reefs Optimization (CRO) algorithm.

Algorithm Step	Pseudo-Code for the CRO Algorithm
1	**Require:** CRO algorithm parameters
2	**Ensure:** An optimal feasible individual (best antenna design)
3	Initialize the algorithm and CRO parameters
4	**for** each iteration of the simulation **do**
5	Update values of CRO parameters: predation probability, etc.
6	Broadcast spawning and Brooding operators
7	Settlement of new corals
8	Predation process
9	Evaluate the new population in the coral reef
10	**end for**
11	**Return:** the best individual (final solution) from the reef

The CRO-SL (Coral Reef Optimization with Substrate Layers) is an improved version of the CRO [19]. It consists of a multi-method ensemble for optimization [27], with extremely good search capabilities for optimization tasks. The CRO-SL has the same algorithmic structure than the basic CRO, but several substrate layers are defined in the algorithm, each one implementing a different search procedure or strategy. In fact, the CRO-SL is an ensemble approach which promotes competitive co-evolution, where each substrate layer may represent different processes (different models, search operators, problem's parameters, etc.), though the multi-method version, in which the substrate layers represent different search operators, has been the most successful version. Details on the overall CRO-SL algorithm and the mechanisms to include substrate layers are well-reported in reference [19]. The main steps of the CRO algorithm have been detailed in Table 2.

3.1. Substrate Layers Implemented

Though different search strategies can be defined at the practitioner's discretion, this work adopts a five-substrate construct of the CRO-SL. They are briefly described below:

1. HS: Mutation using the Harmony Search procedure. Harmony Search [28] is a well known meta-heuristic based on the how a music orchestra improvises a melody. HS substrate controls the generation of new larvae in one of the following ways: (i) with a probability HMCR in $(0, 1)$ (Harmony Memory Considering Rate), the value of a component of the new solution is drawn uniformly from the same values of the component in other corals of the current reef; and (ii) with a probability PAR in $(0, 1)$ (Pitch Adjusting Rate), where small adjustments are applied to the values of the current solution.

2. DE: Differential Evolution algorithm mutation. This substrate is based on the DE algorithm defined in reference [29]. This approach introduces a differential mechanism for exploring the search space. In this case, new larvae are generated by perturbing the current larva by using a vector of differences between two individuals in the population. This perturbation is defined as $x'_i = x^1_i + F(x^2_i - x^3_i)$ (where F stands for a weighting the perturbation amplitude, 0.6 in this case). After this perturbation of the current larva, the perturbed vector x' is in turn combined with an alternative (different) coral in the reef, by means of a classical 2-points crossover, as defined next.

3. 2Px: Classical two-points crossover. The crossover operator is the most used operator for exploring the search space in evolutionary computation algorithms [30]. It consists of coupling two individuals at random, and then, after choosing two points for the crossover, interchanging the genetic material in between these two points. In the current CRO-SL implementation, one larva to be crossed comes from the 2Px substrate, whereas the other can be chosen from any part of the reef.

4. GM. Gaussian Mutation. We consider a traditional Gaussian mutation of the form $x'_i = x_i + N_i(0, \sigma^2)$, where $N_i(0, \sigma^2)$ is a random number following the Gaussian distribution of 0 mean and variance σ^2. We introduce a linear decreasing of σ value during the algorithm, from 0.2(A-B) to 0.02(A-B), where [B,A] is the domain search. Note that this procedure produces a stronger mutation in the beginning of the algorithm, and a fine tuning of the search with smaller displacements nearing the end or the algorithm's evolution.

5. SAbM: Strange Attractors-based Mutation. This is a new search operator proposed in reference [31], specifically designed to use fractal geometric patterns in the search of new larvae. Specifically, it is designed to generate structures of non-linear dynamical systems with chaotic behavior [32]. Interested reader may consult reference [31] to obtain more information on this operator.

3.2. Objective Function: Antenna Simulation and Calculation

The objective function considered $(f(x))$ to guide the algorithm toward optimal antenna optimization and takes into account different design requirements of the device, such as its resonant frequency and bandwidth. Specifically, in order to calculate $f(x)$, we first take into account a discretization of the S_{11} antenna parameter, which is calculated by simulation using the CST software, as described in the next subsection. In this case, a discretization in steps of 2 MHz is considered. To calculate $f(x)$, several frequency bands for mobile communication systems (including LTE and 5G) has been considered. For each frequency band, the mathematical formulation of the objective function is the following:

$$g_f(x) = 0.8 \cdot N^{-10dB} + 0.1 \cdot M + 0.1 \cdot M^* \tag{1}$$

where N^{-10dB} stands for the number of S_{11} points in the observation window under -10 dB, $M =$ | mean (S_{11})|, $M^* = |\min(S_{11})|$ and f stands for a given selected frequency band. The final objective function value $f(x)$ is obtained by adding the value of $g_f(x)$ for all the frequency bands f considered:

$$f(x) = \sum_f g_f(x) \tag{2}$$

Note that this function takes into account the antenna bandwidth for all the frequency bands considered, the resonant frequency, and the actual value of the antenna reflection coefficient. In order to make the calculation, we define a measurement window at each selected frequency band, with a resolution of 20MHz: (1) $f_1 = 791-870$ MHz (5G); (2) $f_2 = 1.7-2.3$ GHz (LTE); (3) $f_3 = 3.3-3.8$ GHz (5G). Note that these frequency bands cover the majority of communications services such as 2G/3G/4G and also LTE and 5G bands of ultimate mobile communication systems.

Antenna Simulation with CST Software

CST Microwave Studio© from Dassault Systèmes SE (France) is a well-known software package for electromagnetic analysis and simulation in the high frequency range. It is able to provide a fully automatic meshing procedure of any electromagnetic device and its simulation using different possible simulation techniques, depending on the case (transient solver, frequency domain solver, integral equation solver, etc.). The idea is to launch the antenna simulation for each solution encoded in the CRO-SL algorithm (a given antenna design). The CRO-SL has been coded in Matlab, and we are able to call the CST simulation directly from Matlab, so the hybridization of the CRO-SL and the CST software for antenna simulation is direct. Figure 2 shows this hybridization process: the CST software is launched in order to simulate each larva (antenna design) in the CRO-SL. After the antenna simulation, we calculate the value of the objective function associated with the simulated antenna using Equations (1) and (2), and this value is used in the evolution of the CRO-SL algorithm.

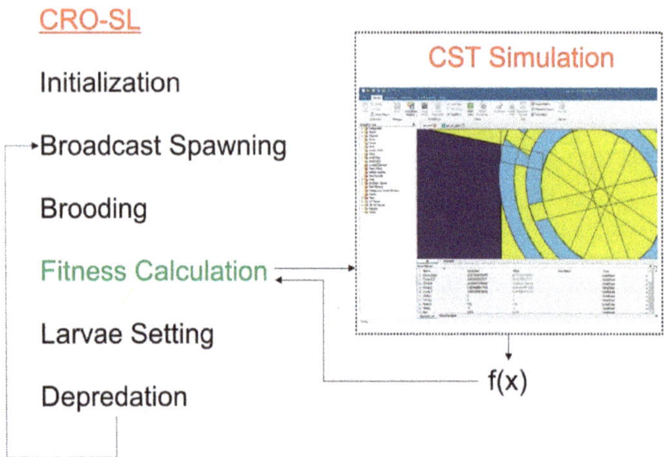

Figure 2. Hybridization of the CST simulation software with the Coral Reefs Optimization with Substrate Layer (CRO-SL) for optimized the parameters of the proposed antenna.

4. Computational Evaluation and Results

This section presents the computational evaluation of the proposed antenna, by means of several simulations using the CRO-SL and CST software. In the first experiment carried out, we consider the highest two frequency bands (f_2 and f_3, associated with LTE and 5G mobile communication systems). In this case, the CRO-SL algorithm is able to obtain a good solution for the problem, within 50 generations, as can be in the S_{11} antenna parameter obtained (Figure 3), with peaks under -10 dBs in both frequency bands considered. The best solution (antenna) obtained by the CRO-SL is shown in Figure 4 (front and bottom view). In this case, the fitness evolution followed by the CRO-SL is shown in Figure 5. The CRO-SL performance depends on how the different substrates operate for this problem. Figure 6 shows the percentage of best solutions and the number of solutions got into the reef for the different substrates. This is a good indication of the best substrates for this specific optimization problem. In this case, the GM substrate is the most active substrate in the CRO-SL to obtain good solutions for the problem, followed by the SAbM and 2Px operators.

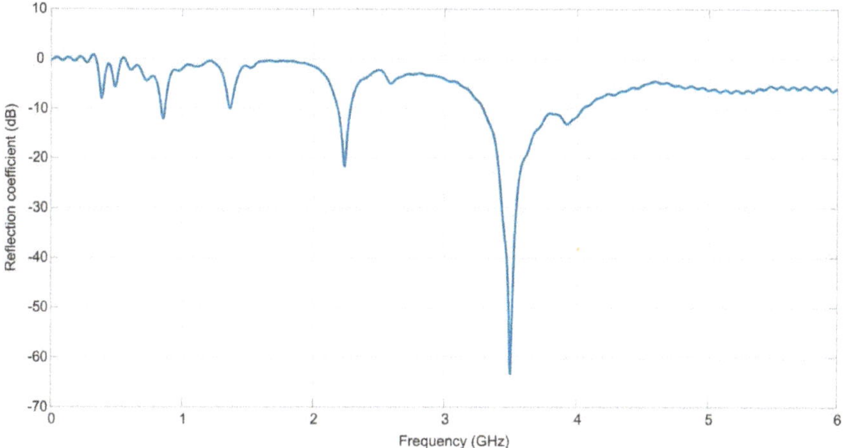

Figure 3. Reflection coefficient of the proposed antenna optimized for f_2 and f_3 frequency bands.

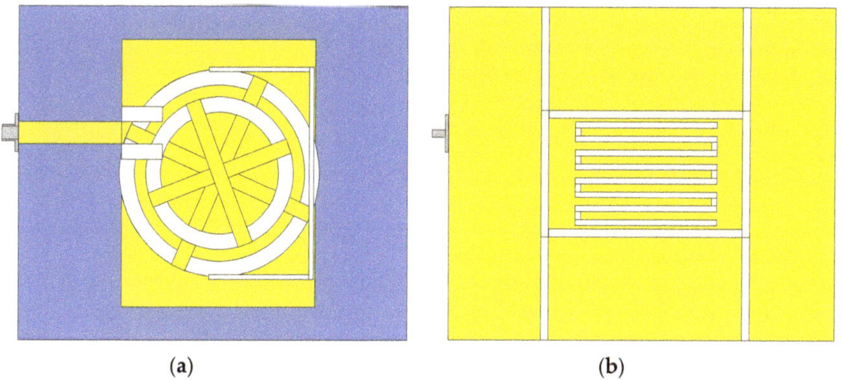

(a)   (b)

Figure 4. Final antenna layout for the two-frequency bands optimization case, after the optimization process. (**a**) Front view; (**b**) bottom view.

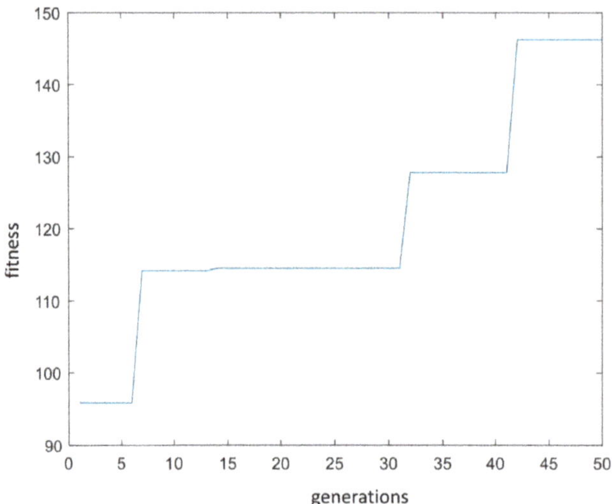

Figure 5. Evolution of the CRO-SL fitness (Equation (2)) in two frequency bands antenna optimization process.

The design of the antenna with the CRO-SL considering the three frequency bands for LTE and 5G has been carried out using the solution for the two frequencies shown above as initial point. In order to do this, we include the best solution for the two-frequencies case into the initial population of the CRO-SL, completing it with randomly-generated solutions and a number of variations of the two-frequencies case obtained by mutation of the best solution. With this, the evolution of the CRO-SL towards a high-quality antenna, able to respond in the three frequency bands considered was really fast. Figure 7 shows the S_{11} antenna parameter obtained by the CRO-SL. Note that the solution obtained is extremely good. In Figure 7a, it is possible to visualize three peaks under -25 dBs in the three frequency bands, with a peak under -45 dB in the 5G frequency centered in 3.5 GHz, and good bandwidth associated with all the frequencies considered. In the same way, the results represented in Figure 7b confirm the good performance of the optimized antenna. Please note that the markers on Figure 7b show the center frequency for each service, but not the exact resonant frequency.

To further evaluate the performance of the CRO-SL algorithm, the optimization problem with three bands has been also tackled with an Evolutionary Algorithm. The initial parameters of the EA are the same as the CRO (population size and number of iterations) in order to be fully comparable. Table 3 shows the comparison between the CRO-SL algorithms and an Evolutionary algorithm. This table shows the best objective function obtained by each compared algorithm (CRO-5SL optimizing two bands, CRO-5SL optimizing three bands, and EA optimizing three bands). It can be seen that the CRO reach a better solution than the EA even when it is optimizing juts two frequency bands.

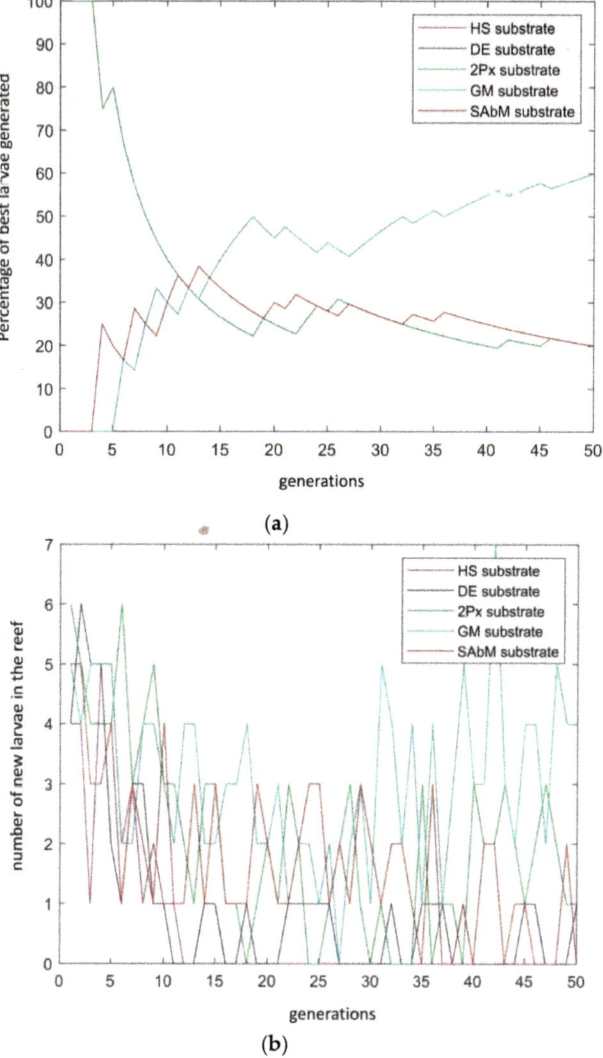

Figure 6. CRO-SL substrate performance metrics for the two-frequencies antenna optimization. (**a**) Best substrate for larvae generation; (**b**) best substrate for getting larvae into the reef.

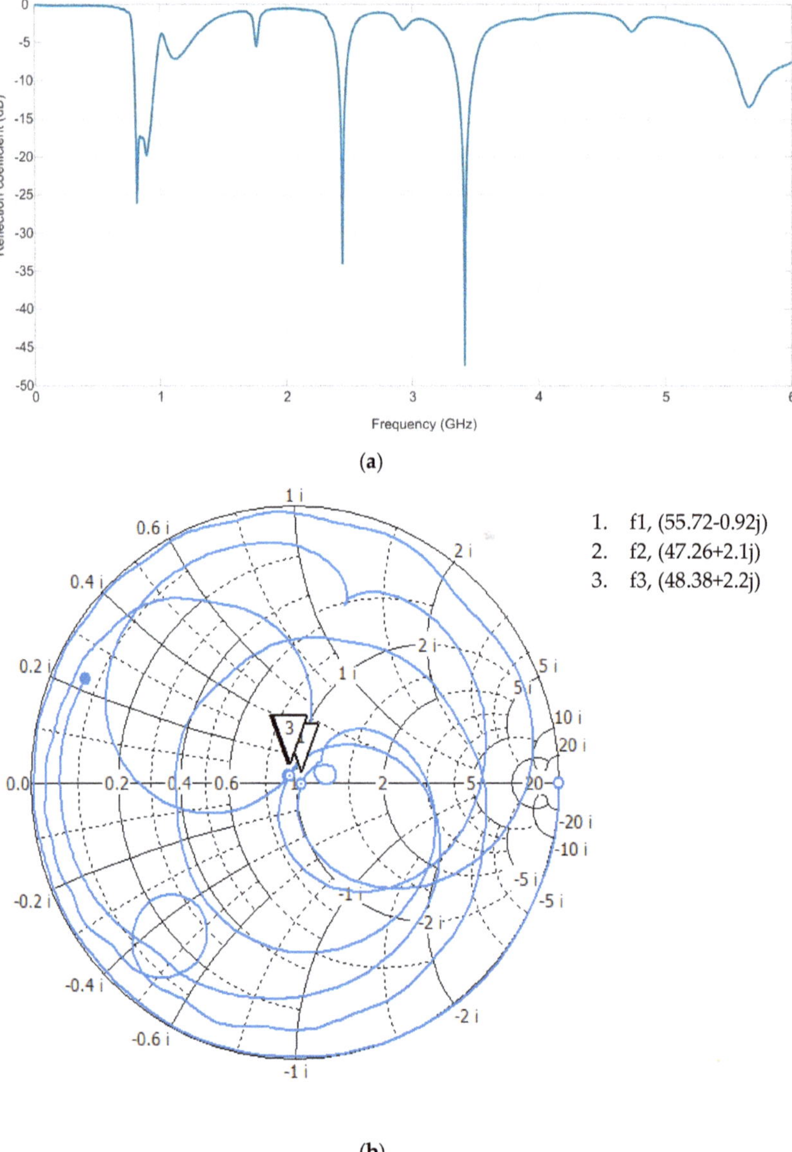

Figure 7. Reflection coefficient of the proposed antenna optimized for f_1, f_2 and f_3 frequency bands. (a) Reflection coefficient in dB; (b) reflection coefficient in Smith Chart.

Table 3. Comparison of the best results obtained by the proposed CRO-SL approaches and an evolutionary algorithm.

Algorithm	Best Fitness
CRO-SL (two frequency bands)	146.24
CRO-SL (three frequency bands)	155.03
Evolutionary Algorithm	130.05

Figure 8 shows the final antenna layout obtained with the CRO-SL algorithm when the three frequency bands are considered. As can be seen after a comparison with the two-frequency bands case, the obtained antenna for the three frequency band shows a more reduced Circle R characteristic, with a wider U_T. There are also differences in the back-side of the optimized antenna (meanders design) when compared to the two-frequencies case.

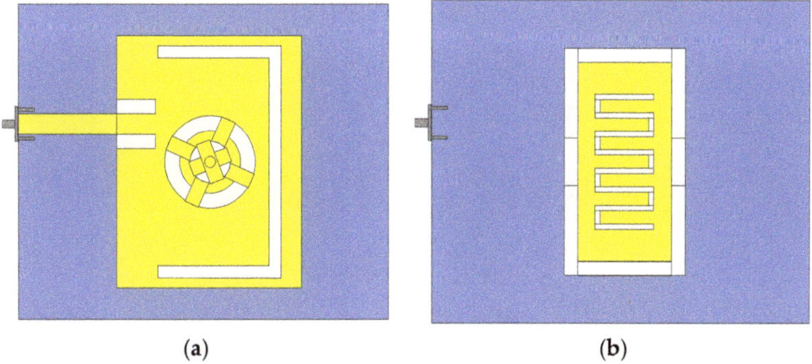

(a)   (b)

Figure 8. Final antenna layout for the three-frequency bands optimization case, after the optimization process; (**a**) Front view; (**b**) Bottom view.

Figure 9 shows the surface current distribution of the antenna for 800, 2400, and 3500 MHz bands. As can be seen, the surface current is concentrated around the U slot for the lower band. In the upper band, the higher values are concentrated around the square patch. The current distribution in the middle band is mainly located in the feeding line and the annular patch. The influence of the ground plane meanders and slots is also shown in the figure.

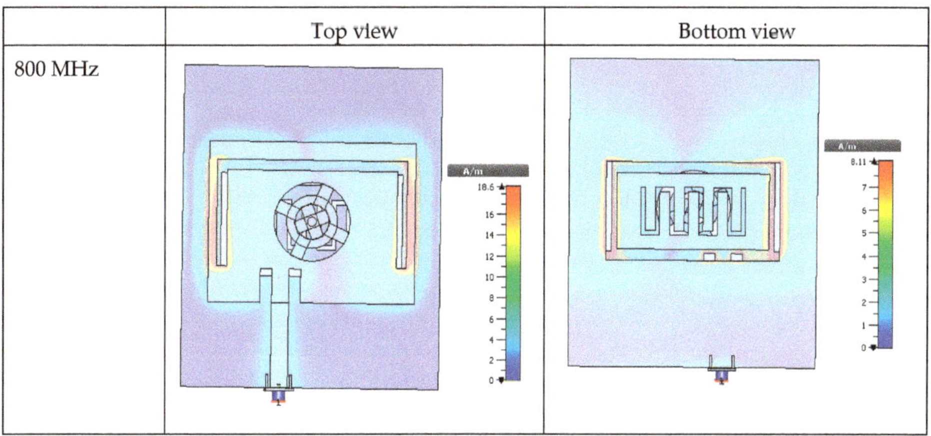

Figure 9. *Cont.*

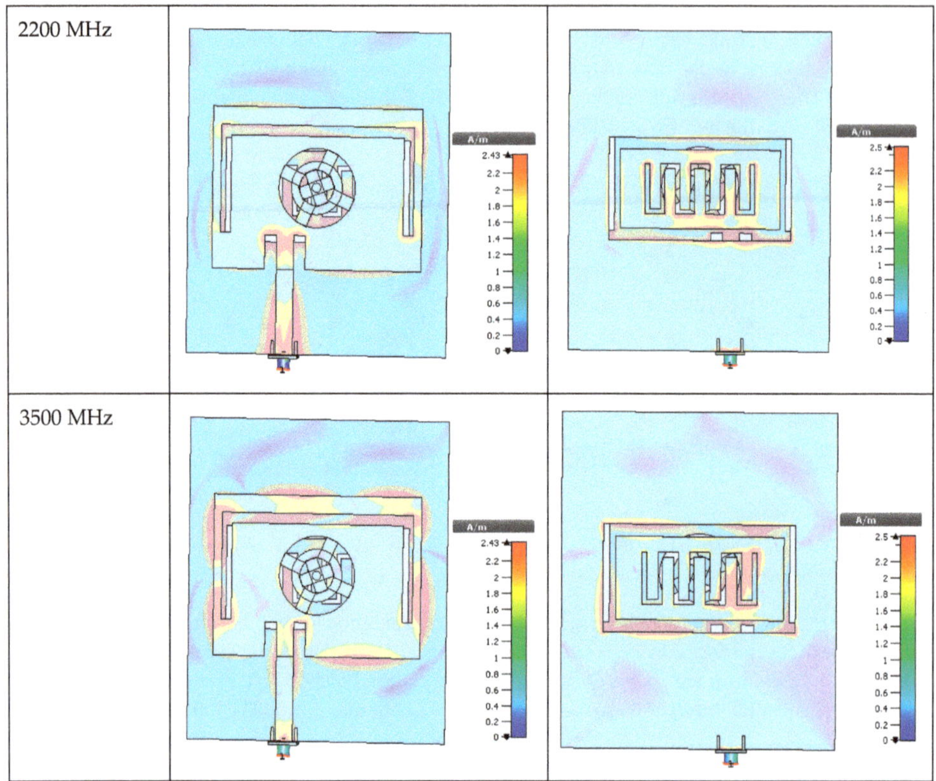

Figure 9. Surface current distribution for the antenna obtained in the three-frequency bands optimization case.

Radiation patterns and gain values for the antenna are given in Figure 10 and Table 4. At 800 MHz, the results are similar to the basic U slot antenna, as the current distribution is mainly located around it and also, the meanders and slots in the ground plane do not affect specially in this case. For the rest of the cases, the radiation pattern differs from the basic equivalent antennas, i.e., annular patch or square patch because neither a single element is involved nor the ground plane influence is negligible.

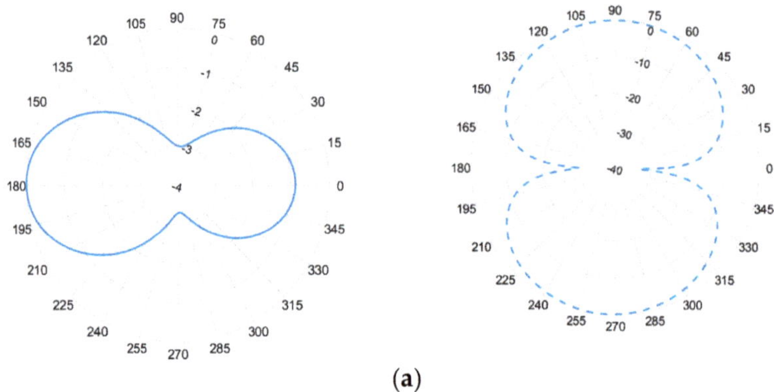

(a)

Figure 10. *Cont.*

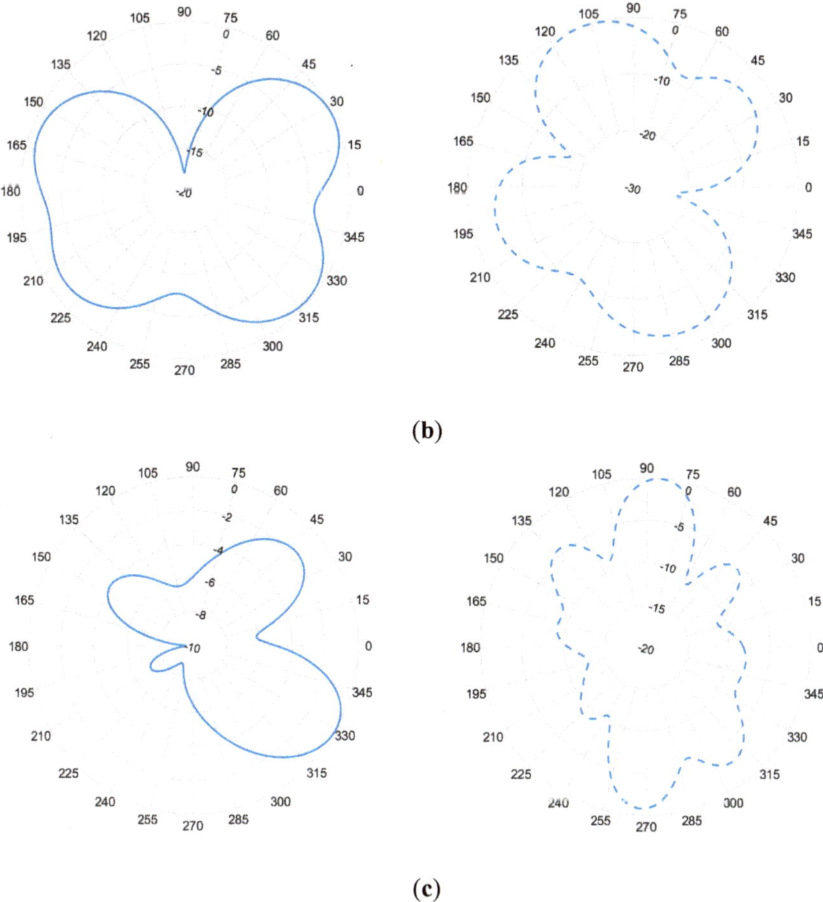

(b)

(c)

Figure 10. Final antenna radiation pattern; (**a**) plane E and plane H at 800 MHz; (**b**) plane E and plane H at 2.2 GHz; (**c**) plane E and plane H at 3.5 GHz.

Table 4. Gain (dBi) for the antenna obtained in the three-frequency bands optimization case.

Frequency (GHz)	Gain (dBi)
0.8	2.734
2.2	4.793
3.5	8.344

The performance of the CRO-SL algorithm in this problem is shown in Figure 11 (fitness evolution), and Figure 12 performance metrics for the different substrates of the CRO-SL. Again, the GM is the most active operator in improving the searching capabilities of the algorithm. The SAbM and the 2Px are also important in the last stages of the algorithm, both in generating good larvae and when the inclusion of the solutions into the reef is considered.

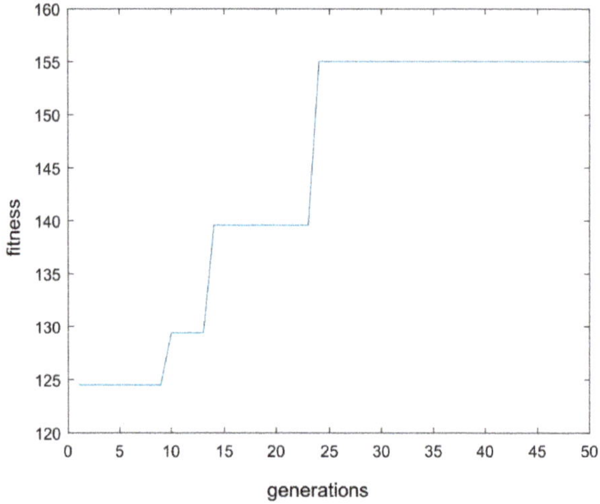

Figure 11. Evolution of the CRO-SL fitness (Equation (2)) in the three-frequency bands antenna optimization process.

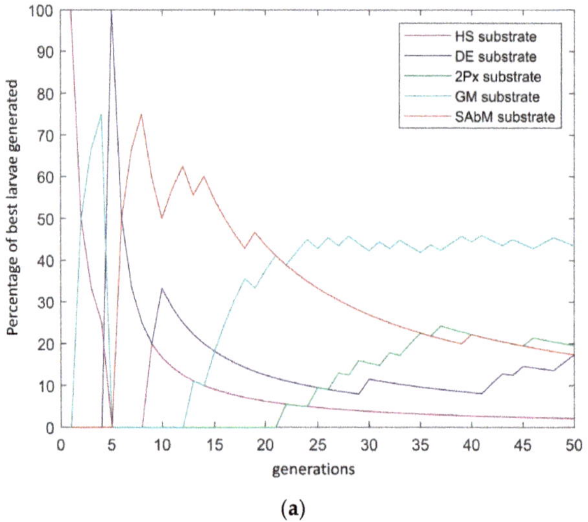

(**a**)

Figure 12. *Cont.*

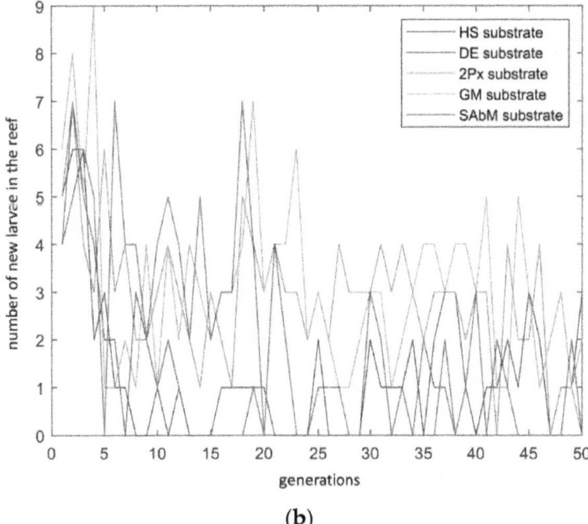

(b)

Figure 12. CRO-SL substrate-performance metrics for the two-frequencies antenna optimization; (a) Best substrate for larvae generation; (b) Best substrate for getting larvae into the reef.

Here, we have shown how we are able to optimize a multi-band microstrip textile patch antenna, with capabilities in 5G services, with the CRO-SL algorithm. The obtained device after the optimization is a robust antenna, able to operate in the three frequency bands considered (850 MHz, 2.2 GHz and 3.5 GHz), where communication services such as 5G operate. The results shown for the two-frequencies and three-frequencies cases optimization shown that we are able to obtain a small textile antenna with extremely good capabilities in the desired frequency bands.

5. Conclusions

In this paper, we have presented the optimization of a textile multi-band antenna, considering LTE and 5G frequency bands. The optimized device is composed by a rectangular microstrip patch with two concentric annular slots, a U-Shaped slot, and three sleeved meanders in the ground plane. This novel form makes the antenna optimization very hard, since its layout depends on a number of design variables, and the antenna must be simulated in order to obtain an optimization result. Thus, traditional optimization methods are not applicable, and meta-heuristics algorithms are the best option. In this case, we propose a novel ensemble-based algorithm—the CRO-SL, a kind of evolutionary algorithm able to combine different search operators within a single population of possible solutions to the problem. The objective function which guides the evolution of the algorithm is obtained after the antenna simulation with a specific simulation software for electromagnetic analysis in the high frequency range. In the experimental evaluation of the proposed method, we have shown the design of two different kind of antennas for two and three frequency bands, both including 5G frequencies. In both cases, the CRO-SL algorithm is able to obtain designs with a high performance in the required frequency bands, as shown in the S11 antenna parameter obtained. The textile antennas designed have a small size (around 60 mm), and the antenna substrate would allow their inclusion in different types of fabrics or clothes, providing them with LTE and 5G capacity connection.

Author Contributions: R.S.-M., D.M.-V., and P.-L.L.-E. performed the simulations, built the platform to connect the software to do the optimization with the electromagnetic simulation tool. C.C.-G. and S.S.-S. designed and adjusted the optimization algorithm. All the authors have contributed to writing and reviewing the paper. All authors have read and agreed to the published version of the manuscript.

Funding: This work has been partially supported by the Spanish Ministerial Commission of Science and Technology (MICYT) through project number TIN2017-85887-C2-2-P.

Conflicts of Interest: The authors declare no conflict of interest.

References

1. Panwar, N.; Sharma, S.; Singh, A.K. A survey on 5G: The next generation of mobile communication. *Phys. Commun.* **2016**, *18*, 64–84. [CrossRef]
2. Locher, I.; Klemm, M.; Kirstein, T. Troster. Design and characterization of purely textile patch antennas. *IEEE Trans. Adv. Packag.* **2006**, *29*, 777–788. [CrossRef]
3. Hertleer, C.; Hendrik, R.; Vallozzi, L.; van Langenhove, L. A textile antenna for off-body communication integrated into protective clothing for firefighters. *IEEE Trans. Antennas Propag.* **2009**, *57*, 919–925. [CrossRef]
4. Dierck, A.; Agneessens, S.; Declercq, F.; Spinnewyn, B.; Stockman, G.J.; Van Torre, P.; Vallozzi, L.; Ginste, D.V.; Vervust, T.; Vanfleteren, J.; et al. Active textile antennas in professional garments for sensing, localization and communication. *Int. J. Microw. Wirel. Technol.* **2014**, *6*, 331–341. [CrossRef]
5. Lemey, S.; Agneessens, S.; van Torre, P.; Baes, K.; Vanfleteren, J.; Rogier, H. Wearable flexible lightweight modular RFID tag with integrated energy harvester. *IEEE Trans. Microw. Theory Tech.* **2016**, *64*, 2304–2314. [CrossRef]
6. Bayram, Y.; Zhou, Y.; Shim, B.S.; Xu, S.; Zhu, J.; Kotov, N.A.; Volakis, J.L. E-textile conductors and polymer composites for conformal lightweight antennas. *IEEE Trans. Antennas Propag.* **2010**, *58*, 2732–2736. [CrossRef]
7. Tak, J.; Lee, S.; Choi, J. All-textile higher order mode circular patch antenna for on-body to on-body communications. *Microw. Antennas Propag.* **2015**, *9*, 576–584. [CrossRef]
8. Liu, F.-X.; Kaufmann, T.; Xu, Z.; Fumeaux, C. Wearable applications of quarter-wave patch and half-mode cavity antennas. *IEEE Antennas Wirel. Propag. Lett.* **2015**, *14*, 1478–1481. [CrossRef]
9. Lee, H.; Jinpil, T.; Jaehoon, C. Wearable Antenna Integrated into Military Berets for Indoor/Outdoor Positioning System. *IEEE Antennas Wirel. Propag. Lett.* **2017**, *16*, 1919–1922. [CrossRef]
10. Sanchez-Montero, R.; Salcedo-Sanz, S.; Portilla-Figueras, J.A.; Langley, R. Hybrid PIFA-patch antenna optimized by evolutionary programming. *Prog. Electromagn. Res.* **2010**, *108*, 221–234. [CrossRef]
11. Das, A.; Mandal, D.; Ghoshal, S.P.; Kar, R. Concentric circular antenna array synthesis for side lobe suppression using moth flame optimization. *AEU Int. J. Electron. Commun.* **2018**, *86*, 177–184. [CrossRef]
12. Ustun, D.; Akdagli, A. Design of band notched UWB antenna using a hybrid optimization based on ABC and DE algorithms. *AEU Int. J. Electron. Commun.* **2018**, *87*, 10–21. [CrossRef]
13. Sanchez-Montero, R.; Camacho-Gomez, C.; Lopez-Espi, P.L.; Salcedo-Sanz, S. Optimal Design of a Planar Textile Antenna for Industrial Scientific Medical (ISM) 2.4 GHz Wireless Body Area Networks (WBAN) with the CRO-SL Algorithm. *Sens. J.* **2018**, *18*, 1982. [CrossRef]
14. Awl, H.N.; Abdulkarim, Y.I.; Deng, L.; Bakir, M.; Muhammadsharif, F.F.; Karaaslan, M.; Unal, E.; Luo, H. Bandwidth Improvement in Bow-Tie Microstrip Antennas: The Effect of Substrate Type and Design Dimensions. *Appl. Sci.* **2020**, *10*, 504. [CrossRef]
15. Prabhaka, H.V.; Kummuri, U.K.; Yadahalli, R.M.; Munnappa, V. Effect of various meandering slots in rectangular microstrip antenna ground plane for compact broadband operation. *Electron. Lett.* **2007**, *43*, 848–850. [CrossRef]
16. Hossa, R.; Byndas, A.; Bialkowski, M.E. Improvement of compact terminal antenna performance by incorporating open-end slots in ground plane. *IEEE Microw. Wirel. Compon. Lett.* **2004**, *14*, 283–285. [CrossRef]
17. Zhang, Z.Y.; Fu, G.; Zuo, S.L. A miniature sleeve meander antenna for TPMS application. *J. Electromagn. Waves Appl.* **2009**, *23*, 1835–1842. [CrossRef]
18. Salcedo-Sanz, S.; del Ser, J.; Landa-Torres, I.; Gil-Lopez, S.; Portilla-Figueras, J.A. The Coral Reefs Optimization algorithm: A novel metaheuristic for efficiently solving optimization problems. *Sci. World J.* **2014**, 739768. [CrossRef]
19. Salcedo-Sanz, S. A review on the coral reefs optimization algorithm: New development lines and current applications. *Prog. Artif. Intell.* **2017**, *6*, 1–15. [CrossRef]
20. Langley, R.; Voudouris, K.; Batchelor, J.C. Annular ring patch antennas. In Proceedings of the IEEE Colloquium on Multi-Band Antennas, London, UK, 26 October 1992; pp. 6–11.
21. Mahmoud, S.F. A new miniaturized annular ring patch resonator partially loaded by a metamaterial ring with negative permeability and permittivity. *IEEE Antennas Wirel. Propag. Lett.* **2004**, *3*, 19–22. [CrossRef]

22. Bao, X.L.; Ammann, M.J. Compact annular-ring embedded circular patch antenna with cross-slot ground plane for circular polarization. *Electron. Lett.* **2006**, *42*, 192–193. [CrossRef]
23. Bao, X.L.; Ammann, M.J. Dual-frequency circularly-polarized patch antenna with compact size and small frequency ratio. *IEEE Trans. Antennas Propag.* **2007**, *55*, 2104–2107. [CrossRef]
24. Zhang, Y.; Hong, W.; Yu, C.; Kuai, Z.Q.; Don, Y.; Zhou, J.Y. Planar ultrawideband antennas with multiple notched bands based on etched slots on the patch and/or split ring resonators on the feed line. *IEEE Trans. Antennas Propag.* **2008**, *56*, 3063–3068. [CrossRef]
25. Lee, K.F.; Yang, S.L.S.; Kishk, A.A.; Luk, K.M. The versatile U-slot patch antenna. *IEEE Antennas Propag. Mag.* **2010**, *100*, 71–88. [CrossRef]
26. Hsieh, C.; Chiu, T.T.; Lai, C. Compact dual-band slot antenna at the corner of the ground plane. *IEEE Trans. Antennas Propag.* **2009**, *57*, 3423–3426. [CrossRef]
27. Wu, G.; Mallipeddi, R.; Suganthan, P.N. Ensemble strategies for population based optimization algorithms—A survey. *Swarm Evol. Comput.* **2019**, *44*, 695–711. [CrossRef]
28. Geem, Z.W.; Kim, J.H.; Loganathan, G.V. A new heuristic optimization algorithm: Harmony Search. *Simulation* **2001**, *76*, 60–68. [CrossRef]
29. Storn, R.; Price, K. Differential Evolution—A simple and efficient heuristic for global optimization over continuous spaces. *J. Glob. Optim.* **1997**, *11*, 341–359. [CrossRef]
30. Eiben, A.E.; Smith, J.E. *Introduction to Evolutionary Computing*; Springer: Berlin, Germany, 2003.
31. Salcedo-Sanz, S. Modern meta-heuristics based on nonlinear physics processes: A review of models and design procedures. *Phys. Rep.* **2016**, *655*, 1–70. [CrossRef]
32. Grassberger, P.; Procaccia, I. Characterization of strange attractors. *Phys. Rev. Lett.* **1983**, *50*, 346–349. [CrossRef]

Article

Design and Implementation of Quad-Port MIMO Antenna with Dual-Band Elimination Characteristics for Ultra-Wideband Applications

Pawan Kumar [1,*], Shabana Urooj [2,*,†] and Fadwa Alrowais [3]

[1] Department of Electrical Engineering, School of Engineering, Gautam Buddha University,
   Greater Noida 201308, India
[2] Department of Electrical Engineering, College of Engineering, Princess Nourah Bint Abdulrahman
   University, Riyadh 84428, Saudi Arabia
[3] Department of Computer Sciences, College of Computer and Information Sciences,
   Princess Nourah Bint Abdulrahman University, Riyadh 84428, Saudi Arabia; fmalrowais@pnu.edu.sa
* Correspondence: pawangupta.iitb@gmail.com (P.K.); SMUrooj@pnu.edu.sa (S.U.)
† On leave from Gautam Buddha University, Greater Noida, India.

Received: 4 February 2020; Accepted: 24 February 2020; Published: 2 March 2020

Abstract: A planar, microstrip line-fed, quad-port, multiple-input-multiple-output (MIMO) antenna with dual-band rejection features is proposed for ultra-wideband (UWB) applications. The proposed MIMO antenna design consists of four identical octagonal-shaped radiating elements, which are placed orthogonally to each other. The dual-band rejection property (3.5 GHz and 5.5 GHz corresponding to Wi-MAX and WLAN bands) was obtained by introducing a hexagonal-shaped complementary split-ring resonator (HCSRR) in the radiators of the designed antenna. The MIMO antenna was etched on low-cost FR-4 dielectric substrate of size $58 \times 58 \times 0.8$ mm^3. Isolation higher than 18 dB and envelope correlation coefficient (ECC) lesser than 0.07 was observed for the MIMO/diversity antenna in the operating range of 3–16 GHz. The presented four-port UWB MIMO antenna configuration was fabricated, and the experimental results validate the simulation outcomes.

Keywords: antenna; MIMO; octagonal; planar; UWB

---

1. Introduction

The use of ultra-wideband (UWB) technology for wireless applications has seen a surge after the Federal Communications Commission (FCC) specified the 3.1–10.6 GHz band as an unlicensed band [1]. Recently, UWB antennas have attracted the focus of RF engineers and researchers due to their extensive usage in sensing networks, cognitive radios, microwave imaging, wearable devices, military applications, wireless personal area networks, and high data rate communication systems. The monopole-based UWB antennas of various shapes and sizes have been explored by the researchers for various applications [2,3]. The features, such as low-profile, miniature size, light-weight, good radiation efficiency, and simple integration into communication devices, make planar monopole antennas a preferred choice for UWB transceiver systems [4]. However, UWB antennas suffer from the disadvantage that radiation can be transmitted to smaller distances only, which is due to the use of low power for transmission, as mentioned by the Federal Communications Commission (FCC) [5]. The limitations of low power and short-range transmission can be encountered using multiple-input-multiple-output (MIMO) technology along with UWB. It uses multiple radiating elements for transmitting and receiving wireless signals, but the placement of multiple elements in a limited space within a communication device/system is a challenge. The close placement of resonating elements will result in poor isolation and high envelope correlation coefficient (ECC), which deteriorates the functioning of the designed system. An inverse relationship exists between the mutual coupling and the space present among the resonating elements. Various methods have been explored and proposed by the researchers for designing MIMO antennas. In literature, different configurations of dual-element antennas have been reported, but these antennas provide limited diversity. The designing of quad-port MIMO systems having four identical radiating elements is more complex as compared to the designing of dual-port systems having two identical elements, due to a higher level of mutual coupling in quad-port systems [6,7].

Moreover, various commercially available devices use Wi-MAX and WLAN bands for performing operations like data transfer, wireless connectivity, etc., which may interfere with the UWB and thus affects the performance of the UWB system. Thus, it is preferable to eliminate these bands for better functioning of the UWB system [8,9]. In the literature, many band-notched UWB MIMO/diversity antennas have been explored [10–18]. A MIMO antenna with two G-shaped elements was reported, where isolation between the resonating elements was enhanced by a T-shaped metal strip designed on the ground surface [10]. In [11], a compact MIMO design comprised of a coplanar waveguide-fed (CPW) monopole antenna and a half-slot antenna was presented, where a T-shaped parasitic stub, a protracted Z-shaped stub, a rectangular slot, and a combination of C-shaped slots were implanted to notch 5–6 GHz band. In [12], a small-sized two-port MIMO antenna comprised of square-shaped monopole resonators and a T-shaped metal stub embedded in the ground surface to reduce mutual coupling was proposed. In [13], a MIMO antenna composed of two staircase-shaped resonating elements was reported, where band rejection property was realized by engraving split-ring resonator (SRR) slits on the patch elements. In [14], a planar filtenna with shunt short-circuited stubs and three interdigital edge-coupled patterns was suggested for UWB MIMO applications, where dual notched-band characteristics were realized by introducing a complementary split-ring resonator (CSRR). A small-sized differential stepped-slot MIMO antenna was proposed with notched band rejection capability [15], where four U-shaped stubs were implanted in the radiators to notch the WLAN band. In reference [16], a MIMO antenna consisting of four circular monopole patches and electromagnetic band-gap (EBG) elements was suggested, where four semi-circular ring-shaped slots were engraved on the radiating patches to obtain a notched-band. A four-element UWB MIMO antenna design with a mushroom-like EBG structure to a notch 5.5 GHz band was suggested [17]. In [18], a quad-port UWB MIMO structure with two notched frequency bands was suggested, where the dimensions of the antenna were optimized by using an iterative design method.

In this communication, a planar, compact, MIMO antenna comprising four identical octagonal-shaped resonating elements is proposed. The unit element contains a microstrip line of impedance 50 Ω, octagonal-shaped monopole resonator, and a rectangular-shaped ground surface. A hexagonal-shaped complementary split-ring resonator (HCSRR) was embedded in the monopole antenna patch to notch the interfering Wi-MAX and WLAN frequencies. The MIMO/diversity antenna elements are arranged in orthogonal orientation, and spacing was introduced between them to achieve higher isolation. Metallic strips are also inserted between the monopole elements in order to connect their ground surfaces to obtain common voltage intensity in the MIMO structure. The metallic strips also increase the isolation among four resonators of the MIMO/diversity antenna.

2. Antenna Design

2.1. UWB Antenna Element

The diagrammatic design of the proposed unit antenna element is shown in Figure 1. The unit cell consists of a microstrip line-fed octagonal-shaped resonating element and a rectangular ground plane. An HCSRR is implanted in the octagonal patch of the antenna for notching Wi-MAX (3.5 GHz) and WLAN (5.5 GHz) band frequencies. To obtain better impedance matching, a U-shaped slot is carved from the ground plane of the monopole antenna. The antenna is imprinted on the FR-4 dielectric material of 0.8 mm thickness and relative permittivity of 4.4. An ANSYS HFSS® tool is utilized for simulation, designing, and implementation of the designed antenna. The dimension details of the octagonal monopole element are provided in Table 1.

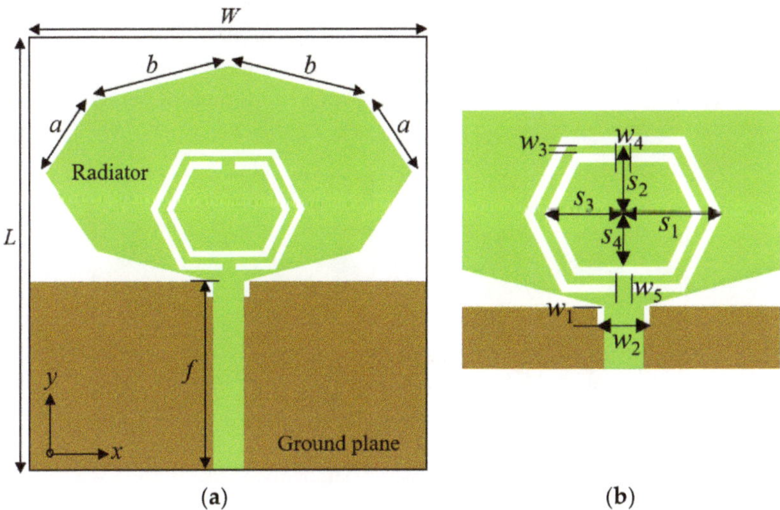

Figure 1. Proposed octagonal-shaped monopole antenna: (**a**) schematic layout; (**b**) enlarged view of the hexagonal-shaped complementary split-ring resonator (HCSRR).

Table 1. Parameter details of the unit element and multiple-input-multiple-output (MIMO) antenna.

Parameter	Dimension (mm)	Parameter	Dimension (mm)
L	26	w_1	1
W	24	s_4	3.5
a	5.7	w_2	2.5
b	8	w_3	0.4
f	11.25	w_4	1
s_1	5	w_5	1
s_2	4.5	L_1	58
s_3	4	w_6	0.5

The antenna design stages are demonstrated in Figure 2. Firstly, an octagonal-shaped radiator (stage-1), fed by a microstrip line of impedance 50 Ω was considered, as displayed in Figure 2a. The radiator was characterized by high eccentricity, which is required for supporting multiple modes in UWB applications. The surface below the antenna radiating patch acted as an unbalanced impedance as it was not grounded. A small U-shaped slot was carved from the ground surface (beneath the microstrip feed line) for providing the required impedance matching among the patch and the feed line (stage-2), as demonstrated in Figure 2b. In Figure 2c, a hexagonal SRR was implanted in the antenna radiator to remove the interfering 5.5 GHz (WLAN) band from the UWB (stage-3). In the next stage, as shown in Figure 2d, an additional hexagonal SRR (opposite to the hexagonal SRR in stage-3) was loaded on the radiator to eliminate interfering 3.5 GHz (Wi-MAX) band signals (stage-4). The lengths of the implanted SRR (outer and inner) can be computed as [19].

$$S_o = 2\pi s_1 - w_5 \approx 0.52\lambda_{go} \tag{1}$$

$$S_i = 2\pi s_3 - w_4 \approx 0.52\lambda_{gi} \tag{2}$$

$$\lambda_g = \frac{c}{f_r}\left(\frac{1}{\sqrt{\varepsilon_{r,\,eff}}}\right) \tag{3}$$

$$\varepsilon_{r,\,eff} = \frac{\varepsilon_r + 1}{2} \tag{4}$$

where λg and f_r denote the guided wavelength and notch-band central frequency, respectively, c denotes the speed of light in vacuum, and $\varepsilon_{r,eff}$ represents the effective relative permittivity. Figure 3 displays the S_{11} characteristics of the design steps of the proposed UWB antenna element. A magnified view of the HCSRR is given in Figure 1b, which contains two concentric hexagonal split rings of different dimensions.

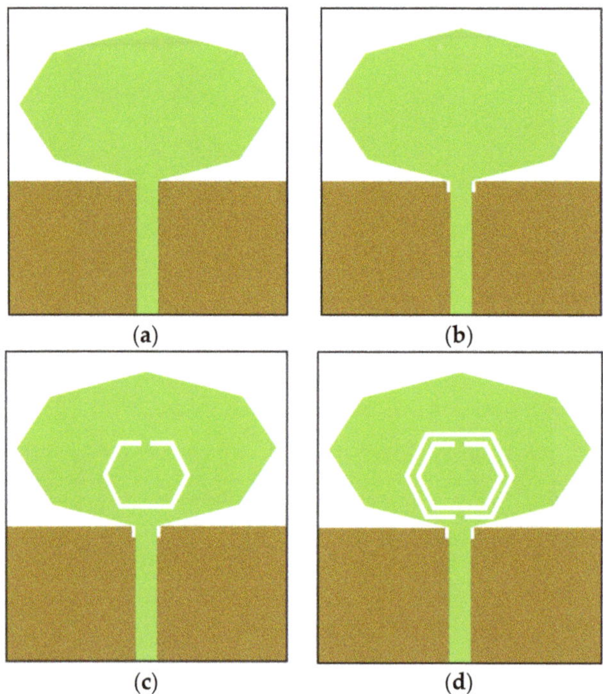

Figure 2. Octagonal-shaped monopole antenna design steps: (**a**) stage-1; (**b**) stage-2; (**c**) stage-3; (**d**) stage-4.

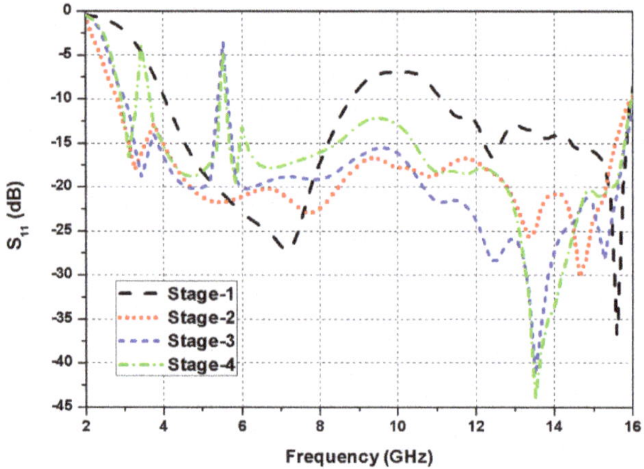

Figure 3. Simulated reflection coefficients comparison of the evolution steps.

Figure 4a,b display the current behavior at notch frequencies 3.5 GHz and 5.5 GHz, correspondingly. From Figure 4a, it can be observed that the concentration of current was higher near the boundaries of the outer hexagonal split-ring, resulting in suppression of the Wi-MAX band. Similarly, Figure 4b shows that a high magnitude current flowed along the boundaries of the inner hexagonal split-ring, which resulted in the elimination of WLAN signals. Therefore, dual band-notched characteristics were achieved by loading an HCSRR in the octagonal radiating patch.

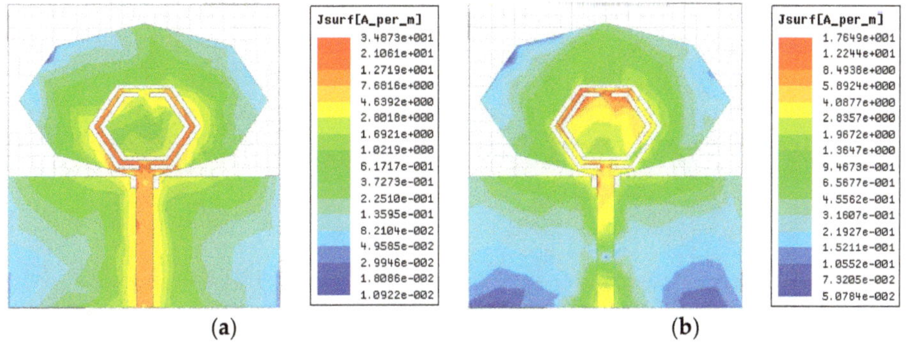

(a)   (b)

Figure 4. Surface current distributions at (**a**) 3.5 GHz; (**b**) 5.5 GHz.

2.2. Quad-Port MIMO Antenna

The geometric layout of the MIMO antenna is presented in Figure 5a, and its design dimensions are provided in Table 1. The MIMO design consists of microstrip-line fed octagonal-shaped four identical antenna elements, which are arranged in orthogonal directions to attain superior performance. The antenna feeding points are represented as port-1, port-2, port-3, and port-4, in the layout diagram. Enhanced isolation was achieved as a result of the orthogonal field produced between the adjacent octagonal-shaped radiating elements, consequently, a reduction in inter element coupling was noticed. However, due to close proximity of the resonating elements, the weak coupling was observed at the lower frequency range, which can be reduced by introducing decoupling elements on the ground surface. Therefore, on the back-side of the dielectric substrate, three metal strips were introduced between the ground patches of the antenna elements to enhance inter-element isolation. The metal strips restrained the current radiated to the adjacent excited elements, thereby enhancing the mutual coupling level. For realizing the optimal decoupling response, the size and spacing among the strips were determined by analyzing multiple simulations and surface current distribution. In addition, these metallic strips united the ground patches of the four radiators to confirm the same voltage in the designed antenna. In a real system, the resonating patches should have the same reference plane to understand the signal level properly. The MIMO system will not work efficiently if isolated ground planes are present [20]. The antenna fabricated prototype is displayed in Figure 5b. The total size of the presented four-port MIMO/diversity antenna is $58 \times 58 \times 0.8$ mm^3.

The proposed MIMO antenna design steps are demonstrated in Figure 6. Firstly, as illustrated in Figure 6a, four identical octagonal-shaped resonating elements were arranged orthogonally to each other. Sufficient space was provided between the four radiators to suppress inter-element correlation. In step-2, as shown in Figure 6b, a thin strip was inserted in the middle of the MIMO antenna to reduce mutual coupling further. The thin strip suppresses the inter-element coupling considerably, but more isolation was required between the antenna elements to obtain effective notching and diversity performance. Therefore, as illustrated in Figure 6c, a decoupling structure consisting of two strips was introduced in the center of the MIMO antenna (step-3), which reduces mutual coupling up to −15 dB in the working band.

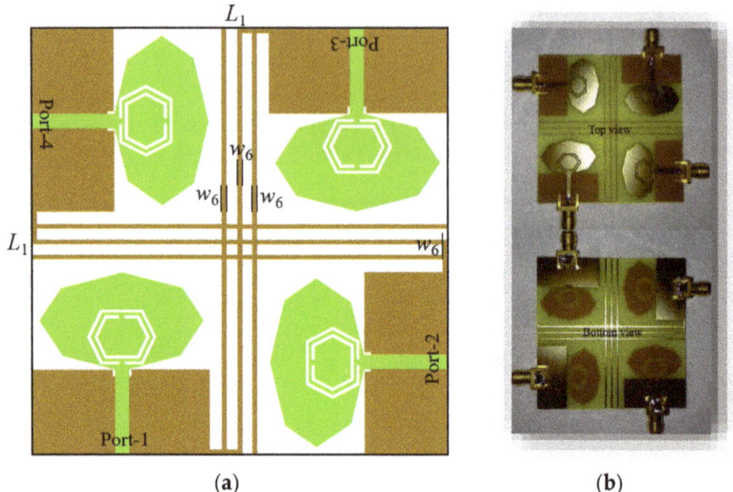

(a)   (b)

Figure 5. Proposed UWB MIMO antenna: (**a**) geometric layout; (**b**) prototype.

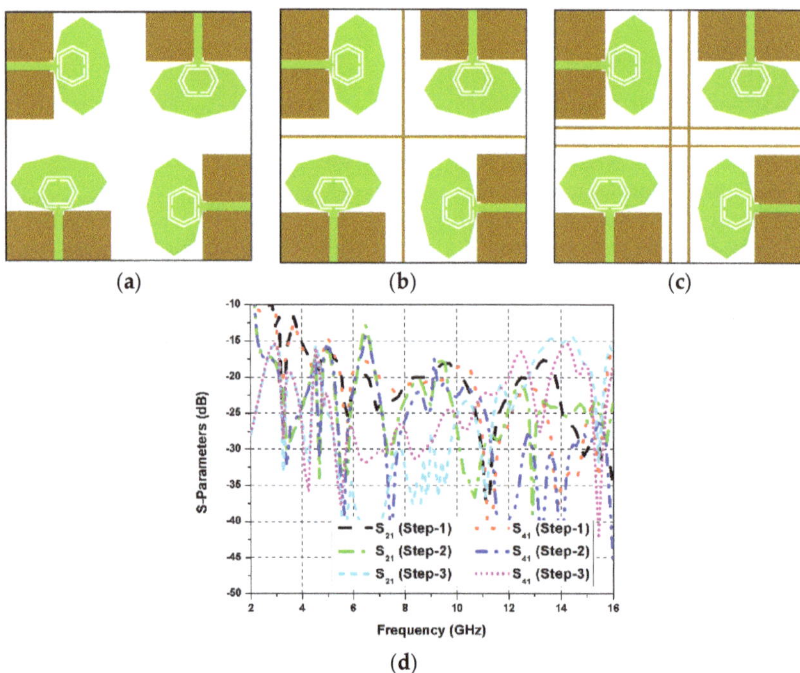

Figure 6. Proposed UWB MIMO antenna design steps: (**a**) step-1; (**b**) step-2; (**c**) step-3; (**d**) mutual coupling comparison.

In addition, to obtain more isolation between the resonating elements, a decoupling element comprised of three metal strips was introduced in the middle of the MIMO antenna (shown in Figure 5). The decoupling element with three metal strips offers isolation greater than 18 dB and also unites the ground planes of the four radiators to confirm the same voltage in the proposed MIMO antenna. The mutual coupling comparison of the three MIMO stages is displayed in Figure 6d.

3. Results

The antenna electrical performance was measured using 50 Ω SMA connectors. The reflection coefficients (simulated and measured) of the designed UWB MIMO antenna at different ports (S_{11}, S_{22}, S_{33}, S_{44}) are shown in Figure 7. While performing measurements at one resonating element, the remaining antenna ports were matched with 50 Ω loads.

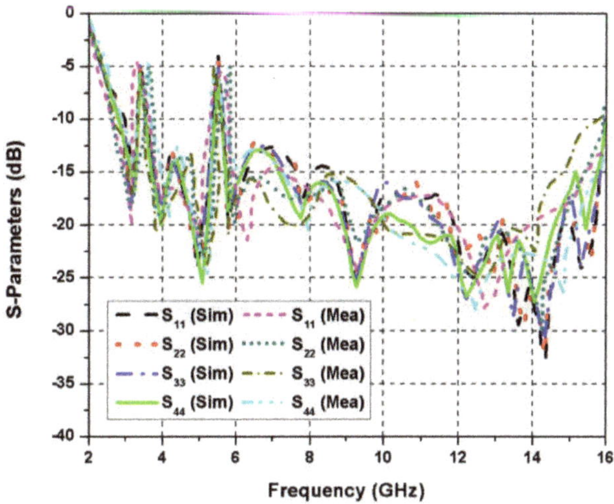

Figure 7. S-parameters (simulated and measured) of the MIMO antenna.

The HCSRR embedded in the antenna radiator eliminated the Wi-MAX and WLAN frequencies. The frequency of the elimination bands can be altered by varying dimensions of the HCSRR. The mutual coupling among different resonating elements of the MIMO antenna is illustrated in Figure 8a,b. The inter-element coupling was decreased through an orthogonal arrangement of the resonating elements and by introducing metal strips between their ground planes. Isolation greater than 18 dB was attained for impedance bandwidth range. As displayed in Figure 9, a 7 dBi peak gain was realized in the proposed UWB antenna. A sudden fall in the gain was observed at dual-band rejection frequencies.

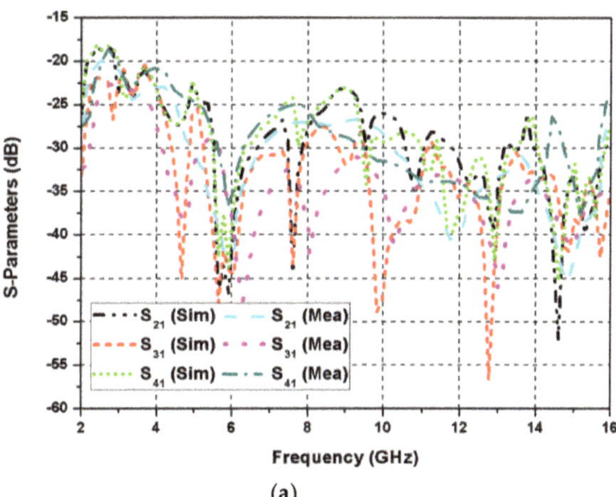

(a)

Figure 8. *Cont.*

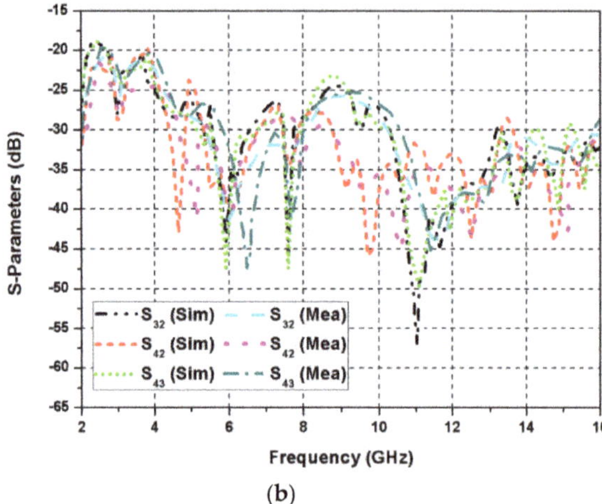

(b)

Figure 8. S-parameters (simulated and measured) of the MIMO antenna at (**a**) port-1; (**b**) other ports.

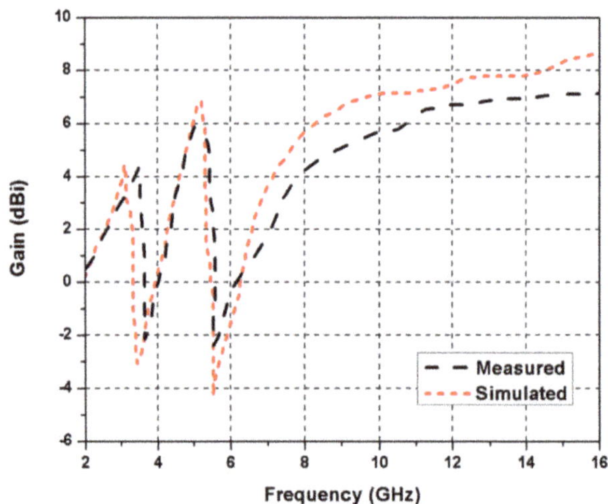

Figure 9. Simulated and measured gain of the proposed antenna.

The surface-current behaviors at notch frequencies 3.5 GHz and 5.5 GHz are presented in Figure 10a,b, correspondingly, here, all the four elements were excited concurrently. The maximum concentration of current (highlighted in red color) was observed near to the boundary of the outer hexagonal split-ring (Figure 10a), which resulted in Wi-MAX band rejection. Similarly, high current concentration near the boundaries of the inner hexagonal split-ring (Figure 10b), marked the rejection of the WLAN band.

ECC can be used to study the coupling effect between adjacent elements of the antenna. The given expression can be used to calculate ECC values between the first and second ports of a multi-port MIMO system [21].

$$\rho_e = \frac{\left|S_{11}^* S_{12} + S_{21}^* S_{22} + S_{13}^* S_{32} + S_{14}^* S_{42}\right|^2}{\left(1 - |S_{11}|^2 - |S_{21}|^2 - |S_{31}|^2 - |S_{41}|^2\right)\left(1 - |S_{12}|^2 - |S_{22}|^2 - |S_{32}|^2 - |S_{42}|^2\right)} \tag{5}$$

In a similar way, the ECC among remaining antenna elements can also be determined. Figure 11 displays the ECC curves among different elements. ECC was observed to remain below 0.07 for the entire UWB band.

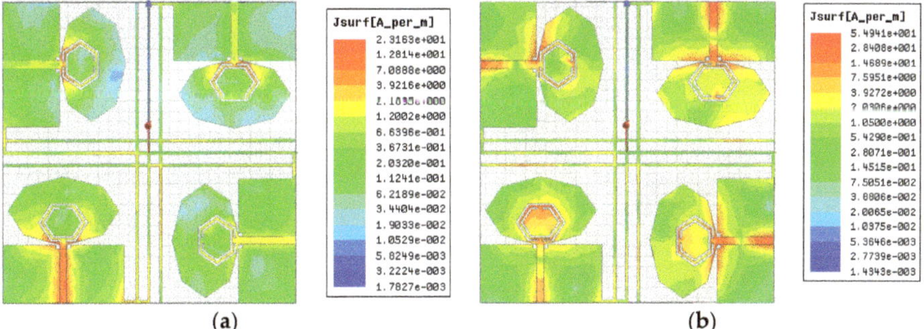

Figure 10. Current distributions at (**a**) 3.5 GHz; (**b**) 5.5 GHz.

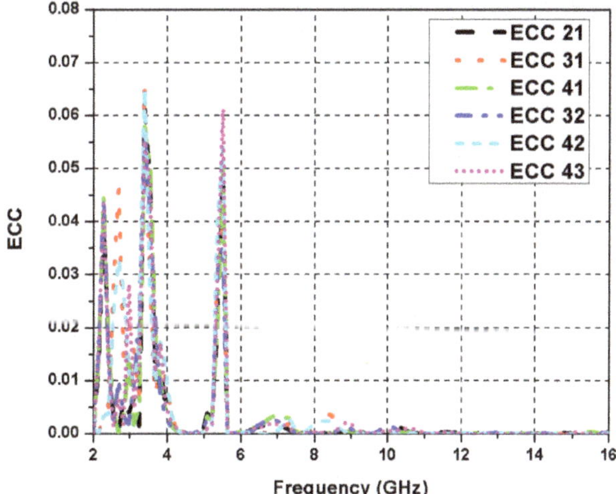

Figure 11. Envelope correlation coefficient (ECC) of the proposed four-port ultra-wideband (UWB) MIMO antenna.

Figure 12 illustrates the co-polar and cross-polar radiation characteristics (simulated and experimental) of the designed MIMO antenna at frequencies 5 GHz, 9 GHz, and 14 GHz. A difference higher than 15 dB was seen among the co-polar and cross-polar pattern curves (in E- and H-planes), which shows the stable functioning of the designed antenna. The co-polar patterns of the E-plane displayed bi-directional behavior while the H-plane co-polar patterns possessed omnidirectional behavior. The simulated and experimental outcomes were found in close proximity. The variations observed between the measured and simulated outcomes were due to high loss tangent, antenna fabrication errors, and soldering of the SMA connectors.

Table 2 lists the comparison of the presented dual-band notched UWB MIMO antenna to other existing UWB MIMO antennas in the literature. The comparison table highlights that the designed antenna has numerous advantages over the previous reported notched-band antennas [10–18], with reference to size, bandwidth ratio, inter-element isolation, and the number of radiating patches. The three metal ring-based decoupling technique used for reducing mutual coupling is simple and effective. In the presented MIMO antenna, the notch frequencies (3.5 GHz and 5.5 GHz) were removed by introducing HCSRR in the resonating patch, without engaging any other active elements or filter circuitry. Moreover, the orthogonal orientation of the resonating elements offered better isolation and joined ground patches of the monopole elements offered a common reference voltage.

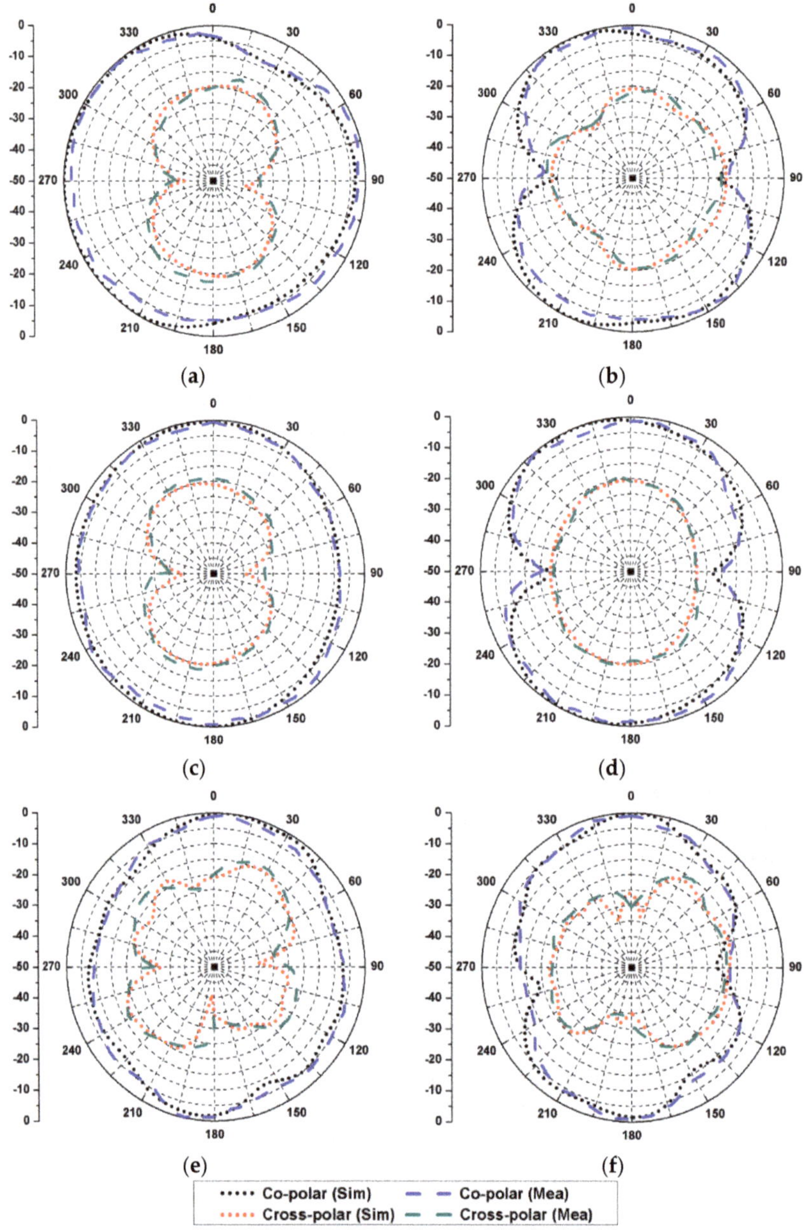

Figure 12. Radiation characteristics of the designed antenna at: (**a**) 5 GHz in H-plane; (**b**) 5 GHz in E-plane; (**c**) 9 GHz in H-plane; (**d**) 9 GHz in E-plane; (**e**) 14 GHz in H-plane; (**f**) 14 GHz in E-plane.

Table 2. Comparison among the presented MIMO antenna and the previously reported band-notched UWB MIMO antennas.

Ref.	Element Number	Bandwidth (GHz)	Size of the Antenna (mm^3)	Notch Band Number	Notched Frequency (GHz)	Isolation (dB)	ECC
[10]	2	2.2–13.35	50 × 82 × 1.6	1	4.4–6.2	>15	<0.04
[11]	2	3.1–10.6	30 × 22 × 0.8	1	5–6	>15	<0.05
[12]	2	3.1–11	22 × 36 × 1.6	1	5.15–5.85	>15	<0.1
[13]	2	2.5–12	48 × 48 × 0.8	1	5.5	>15	<0.005
[14]	2	3.1–10.65	35 × 68 × 1	2	3.35–3.55, 5.65–5.95	>20	<0.002
[15]	4	2.95–10.8	44 × 44 × 1.6	1	5.10–5.95	>15.5	<0.1
[16]	4	3–16.2	60 × 60 × 1.6	1	4.6	>17.5	<0.4
[17]	4	2.73–10.68	60 × 60 × 1.6	1	5.36–6.34	>15	<0.5
[18]	4	2–15	100 × 100 × 1.6	2	3.5, 5.5	>20	<0.1
Prop.	2	3–16	58 × 58 × 0.8	2	3.5, 5.5	>18	<0.07

4. Conclusions

In this article, a small-sized four-port UWB MIMO antenna with dual-band rejection features is presented. The proposed design contains four matching octagonal-shaped resonators arranged orthogonally, and the diagonal elements are arranged anti-parallelly. The antenna showed an isolation >18 dB for the entire operational range. An HCSRR was implanted in the radiating element for notching the interfering Wi-MAX and WLAN frequencies (3.5 and 5.5 GHz, respectively). The simulated and experimental results for gain, S-parameters, isolation, ECC, and radiation patterns were studied. The results validated that the decoupling metal strips used to reduce the inter-element coupling is a simple and efficient approach, and a good diversity response was achieved. The obtained results illustrate that the proposed MIMO antenna could be useful for UWB applications. It can serve as a potential candidate for base station terminals and other wireless communication systems.

Author Contributions: Conceptualization, P.K. and S.U.; methodology, P.K.; software, P.K.; validation, P.K., S.U., and F.A.; formal analysis, P.K.; investigation, P.K.; resources, S.U.; data curation, P.K.; writing—Original draft preparation, P.K.; writing—Review and editing, S.U.; visualization, P.K.; supervision, S.U.; project administration, S.U.; funding acquisition, S.U. and F.A. All authors have read and agreed to the published version of the manuscript.

Acknowledgments: This research was funded by the Deanship of Scientific Research at Princess Nourah Bint Abdulrahman University, Riyadh, Saudi Arabia, through the Fast-Track Research Funding Program.

References

1. Federal Communications Commission. *First Report and Order, Revision of Part 15 of the Commission's Rules Regarding Ultrawideband Transmission Systems*; Federal Communications Commission: Washington, DC, USA, 2002.
2. Chen, Z.N.; Ammann, M.J.; Qing, X.M.; Wu, X.H.; See, T.S.P.; Cai, A. Planar antennas. *IEEE Microw. Mag.* **2006**, *7*, 63–73. [CrossRef]
3. Liang, J.X.; Chian, C.C.; Chen, X.D.; Parini, C.G. Study of a printed circular disc monopole antenna for UWB systems. *IEEE Trans. Antennas Propag.* **2005**, *53*, 3500–3504. [CrossRef]
4. Doddipalli, S.; Kothari, A. Compact UWB antenna with integrated triple notch bands for WBAN applications. *IEEE Access* **2019**, *7*, 183–190. [CrossRef]
5. Nekoogar, F. *Ultra-Wideband Communications: Fundamentals and Applications*; Prentice-Hall: Upper Saddle River, NJ, USA, 2006.
6. Sarkar, D.; Srivastava, K.V. Compact four-element SRR-loaded dual-band MIMO antenna for WLAN/WiMAX/ WiFi/4G-LTE and 5G applications. *Electron. Lett.* **2017**, *53*, 1623–1624. [CrossRef]

7. Ramachandran, A.; Pushpakaran, S.V.; Pezholil, M.; Kesavath, V. A four-port MIMO antenna using concentric square-ring patches loaded with CSRR for high isolation. *IEEE Antennas Wirel. Propag. Lett.* **2016**, *15*, 1196–1199. [CrossRef]

8. Abbas, A.; Hussain, N.; Jeong, M.-J.; Park, J.; Shin, K.S.; Kim, T.; Kim, N. A rectangular notch-band UWB antenna with controllable notched bandwidth and centre frequency. *Sensors* **2020**, *20*, 777. [CrossRef] [PubMed]

9. Rahman, M.; Ko, D.-S.; Park, J.-D. A compact multiple notched ultra-wide band antenna with an analysis of the CSRR-tO-CSRR coupling for portable UWB applications. *Sensors* **2017**, *17*, 2174. [CrossRef] [PubMed]

10. Toktas, A. G-shaped band-notched ultra-widebandMIMO antenna system for mobile terminals. *IET Microw. Antennas Propag.* **2017**, *11*, 718–725. [CrossRef]

11. Tao, J.; Feng, Q. Compact UWB band-notch MIMO antenna with embedded antenna element for improved band notch filtering. *Prog. Electromagn. Res. C* **2016**, *67*, 117–125. [CrossRef]

12. Liu, L.; Cheung, S.W.; Yuk, T.I. Compact MIMO antenna for portable UWB applications with band-notched characteristic. *IEEE Trans. Antennas Propag.* **2015**, *63*, 1917–1924. [CrossRef]

13. Gao, P.; He, S.; Wei, X.; Xu, Z.; Wang, N.; Zheng, Y. Compact printed UWB diversity slot antenna with 5.5-GHz band-notched characteristics. *IEEE Antennas Wirel. Propag. Lett.* **2014**, *13*, 376–379. [CrossRef]

14. Li, W.T.; Hei, Y.Q.; Subbaraman, H.; Shi, X.W.; Chen, R.T. Novel printed filtenna with dual notches and good out-of-band characteristics for UWB-MIMO applications. *IEEE Microw. Wirel. Compon. Lett.* **2016**, *26*, 765–767. [CrossRef]

15. Liu, Y.Y.; Tu, Z.H. Compact differential band-notched stepped-slot UWB-MIMO antenna with common-mode suppression. *IEEE Antennas Wirel. Propag. Lett.* **2017**, *16*, 593–596. [CrossRef]

16. Wu, W.; Yuan, B.; Wu, A. A quad-element UWB-MIMO antenna with band-notch and reduced mutual coupling based on EBG structures. *Int. J. Antennas Propag.* **2018**, *8490740*, 1–10. [CrossRef]

17. Kiem, N.K.; Phuong, H.N.B.; Chien, D.N. Design of compact 4× 4 UWB-MIMO antenna with WLAN band rejection. *Int. J. Antennas Propag.* **2014**, *2014*, 539094. [CrossRef]

18. Shehata, M.; Said, M.S.; Mostafa, H. Dual notched band quad-element MIMO antenna with multitone interference suppression for IR-UWB wireless applications. *IEEE Trans. Antennas Propag.* **2018**, *66*, 5737–5746. [CrossRef]

19. Patre, S.R.; Singh, S.P. Broadband multiple-input-multiple-output antenna using castor leaf-shaped quasi-self-complementary elements. *IET Microw. Antennas Propag.* **2016**, *10*, 1673–1681. [CrossRef]

20. Sharawi, M.S. Current misuses and future prospects for printed multiple-input, multiple-output antenna systems [Wireless Corner]. *IEEE Antennas Propag. Mag.* **2017**, *59*, 162–170. [CrossRef]

21. Blanch, S.; Romeu, J.; Corbella, I. Exact representation of antenna system diversity performance from input parameter description. *Electron. Lett.* **2003**, *39*, 705–707. [CrossRef]

Article

Ultrawideband Low-Profile and Miniaturized Spoof Plasmonic Vivaldi Antenna for Base Station

Li Hui Dai [1], Chong Tan [2] and Yong Jin Zhou [1,*]

[1]   Key Laboratory of Specialty Fiber Optics and Optical Access Networks, Shanghai University,
     Shanghai 200444, China; winsunshine@shu.edu.cn
[2]   Shanghai Institute of Microsystem and Information Technology, Chinese Academy of Sciences,
     Shanghai 200050, China; chong.tan@mail.sim.ac.cn
*   Correspondence: yjzhou@shu.edu.cn

Received: 23 February 2020; Accepted: 31 March 2020; Published: 2 April 2020

Abstract: Stable radiation pattern, high gain, and miniaturization are necessary for the ultra-wideband antennas in the 2G/3G/4G/5G base station applications. Here, an ultrawideband and miniaturized spoof plasmonic antipodal Vivaldi antenna (AVA) is proposed, which is composed of the AVA and the loaded periodic grooves. The designed operating frequency band is from 1.8 GHz to 6 GHz, and the average gain is 7.24 dBi. Furthermore, the measured results show that the radiation patterns of the plasmonic AVA are stable. The measured results are in good agreement with the simulation results.

Keywords: vivaldi antenna; miniaturized; high gain; surface plasmons; ultrawideband

---

1. Introduction

The fifth-generation (5G) mobile communication systems have found various applications, such as autonomous driving [1], telemedicine [2], and virtual reality [3]. Considering radio wave propagation and available bandwidth, the bands between 3 and 5 GHz have been allocated for 5G services in many regions, such as 3.4–3.8 GHz in Europe, 3.7–4.2 GHz in the USA, and 3.3–3.6 and 4.8–4.99 GHz in China [4]. Hence, in the 2G/3G/4G/5G base station applications, the wide bandwidth covers the 2G, 3G, and 4G bands (1.7–2.7 GHz) and the new sub-6 GHz 5G frequency bands. It is urgent to investigate and design ultra-wideband antennas to cover the aforementioned bands and maintain good impedance matching within the entire frequency bands of interest, with a stable half-power bandwidth (HPBW) and a smaller structure.

Crossed dipole antennas are widely used in wireless communication systems due to their advantages of wide bandwidth, stable radiation patterns, and ease of fabrication [5]. Various crossed dipoles were utilized to achieve a wide impedance bandwidth, such as magnetoelectric dipole antennas [6], bow-tie dipole [7], fan-shaped dipole [8], octagonal loop dipole [9], and double-loop dipole [10]. For most of them, the wide impedance bandwidth only covers 2G/3G/4G bands. Furthermore, the antenna profile is not low due to the necessary broadband feeding part.

The microstrip antennas have advantages of low profile, light weight, and ease of conformability. However, their impedance bandwidth and isolation are not good enough. Some attractive techniques have been proposed to broaden the impedance bandwidth by using parasitic patches [11], loading H-slot [12] and U-slot [13], using a multiresonant structure [14], and using a fractal structure [15]. Although the shorted-dipoles printed antenna fed by integrated baluns [16] and the dual-band folded dipole antenna [17] cover 2G/3G/4G/5G frequency band, the gain and radiation pattern of these antennas are unstable over the operating frequency band.

Another typical ultra-wideband antenna is based on tapered slot technique. As one of the most typical types, Vivaldi antenna was firstly introduced in [18] and has found many applications [19,20]. However, its size is still large. Surface plasmons (SPs) are collective electron–photon oscillations

Appl. Sci. **2020**, *10*, 2429; doi:10.3390/app10072429   www.mdpi.com/journal/applsci

confined to the metal surface, which are characterized by subwavelength confinement and field enhancement in optical spectrum [21]. Spoof SPs could be obtained by constructing periodic metal surfaces to get the characteristics of SPs in the microwave frequencies [22,23]. They have been explored in many antenna applications such as holo-graphic antenna [24,25], metasurface planar lenses [26,27], and metasurface superstrate [28,29]. It has been shown that by loading plasmonic metamaterials, the antenna can be efficiently miniaturized [30].

Here, by loading plasmonic metamaterials, a modified antipodal Vivaldi antenna (AVA) is designed and measured, which is composed of the AVA and the loaded periodic slots. The designed operating frequency is from 1.8 GHz to 6 GHz, which covers 2G/3G/4G/5G frequency band. The simulated results and measured results agree well. It shows that the radiation pattern of the plasmonic AVA is stable and the average gain is 7.24 dBi over the operating frequency band.

2. Methods and Principles

The typical monopole antenna and the plasmonic monopole antenna are illustrated in Figure 1a. Usually, the plasmonic monopole antenna is composed of the corrugated monopole antenna with grooves along the metal wire. Here, we use meander wires to simulate the effects of the grooves for simplicity. The dispersion relations of both antennas are calculated and plotted in Figure 1b. When the groove depth b is 0 mm, it is the typical monopole antenna. From Figure 1b, it can be seen that when the groove depth is increased (from 0 mm to 3.6 mm), the corresponding asymptotic frequencies are decreased and the dispersion curves are lowered. It can be concluded that when the wave vector β is fixed, the operating frequency is lower for the plasmonic monopole antenna with deeper grooves. Hence, the electrical size of the plasmonic monopole antenna would be smaller, compared with that of the traditional monopole antenna.

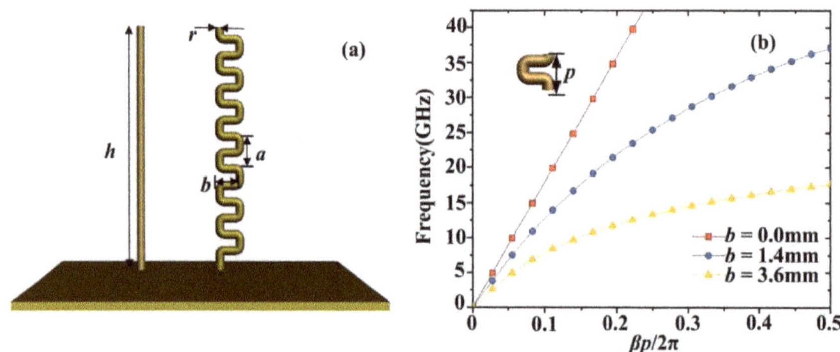

Figure 1. (a) The typical and plasmonic monopole antenna. The dimensions of the antennas: h = 25 mm, a = 0.65 mm, b = 1.4 mm, and r = 0.5 mm. (b) The dispersion relations of both antennas.

The efficiencies of the typical and plasmonic monopole antennas are calculated and illustrated in Figure 2. Figure 2b shows the efficiencies of the plasmonic monopole antenna with different groove depths from 1 mm to 4 mm. From Figure 2b, it can be seen that the operating frequencies red-shifts (corresponding to longer wavelengths) when the groove depth is increased for the fixed height of the monopole antenna (25 mm). Especially, the efficiency curves corresponding to three typical groove depths are plotted in Figure 2c. It can be clearly observed that the central operating frequency red-shifts when the groove depth b is 0 mm (typical monopole antenna), 1.4 mm, and 3.6 mm, respectively. Meanwhile, we can observe that there is an optimized efficiency when tuning the groove depth. Hence, it indicates that the plasmonic theory provides a scheme for miniaturization and optimization of the antenna.

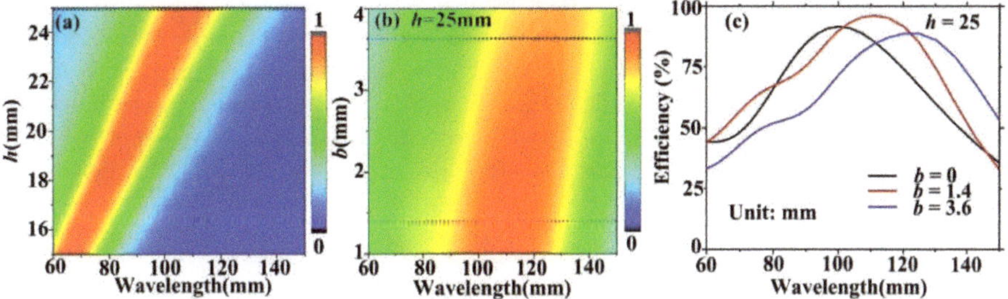

Figure 2. (a) The efficiencies of the typical monopole antenna with different lengths (*h*); (b) the efficiencies of the plasmonic monopole antenna with different groove depths (*h* = 25 mm); and (c) the efficiencies of the plasmonic monopole antenna when the groove depth *b* is 0 mm, 1.4 mm, and 3.6 mm.

3. Design of Plasmonic Antipodal Vivaldi Antenna

The typical AVA is illustrated in Figure 3a, which is denoted by Structure A, where the metal layers are on both sides of the 0.8-mm-thickness F4B substrate ($\varepsilon_r = 2.65$, $tan\delta = 0.001$) with a dimension of 94 mm × 70 mm. A microstrip feeding line is adopted for broadband impendence matching. The port width of the microstrip feeding line is fixed to 1.35 mm. The exponential profile curves employed in this design can be calculated by the following equations:

$$y = c_1 e^{Rx} + c_2 \tag{1}$$

$$c_1 = \frac{y_2 - y_1}{e^{Rx_2} - e^{Rx_1}} \tag{2}$$

$$c_2 = \frac{y_1 e^{Rx_2} - y_2 e^{Rx_1}}{e^{Rx_2} - e^{Rx_1}} \tag{3}$$

The tentative plasmonic AVA is shown in Figure 3b, which is named as Structure B. The final plasmonic AVA is shown in Figure 3c, which is denoted by Structure C, where periodic elliptic grooves are cut from the feeding section to the terminal of the antenna for both structures. Furthermore, the rectangular gradual grooves with a width of 1 mm around the feeding section are further etched to realize the miniaturization of the AVA for Structure C. The detailed structures of the gradual grooves are shown in the insets of Figure 3b,c.

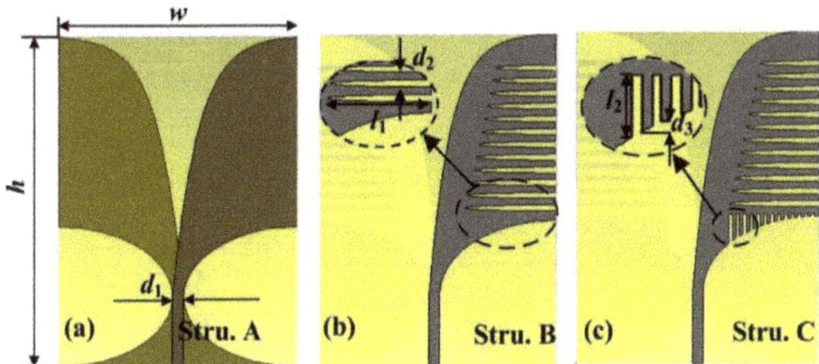

Figure 3. (a) The typical antipodal Vivaldi antenna (AVA). (b) The tentative plasmonic AVA. (c) The final plasmonic AVA. The dimensions of the antenna: $R = 0.16$, $h = 94$ mm, $w = 70$ mm, $l_1 = 26$ mm, $d_1 = 2.5$ mm, $d_2 = 1.25$ mm, $l_2 = 8.02$ mm, and $d_3 = 2$ mm.

The numerical simulations were performed by using CST Microwave Studio. The simulated return losses changing with the grooves are shown in Figure 4. First, from Figure 4a, it can be observed that the impedance matching at lower frequencies is better for Structure C, compared with Structure

B, and the operating frequency red-shifts. Thus, loading the rectangular gradual grooves can make the AVA further miniaturized. Then, the typical parameters of structures B and C are the maximum groove depth l_1 and the groove gap d_2. It can be seen that the operating frequency red-shifts when the maximum groove depth l_1 is bigger, while the groove gap d_2 mainly affects the input impedance at higher frequency. The optimized maximum groove depth l_1 and the groove gap d_2 are set to be 26 mm and 1.25 mm, respectively, to make sure that the lowest operating frequency is 1.8 GHz when the S_{11} is not larger than -10 dB.

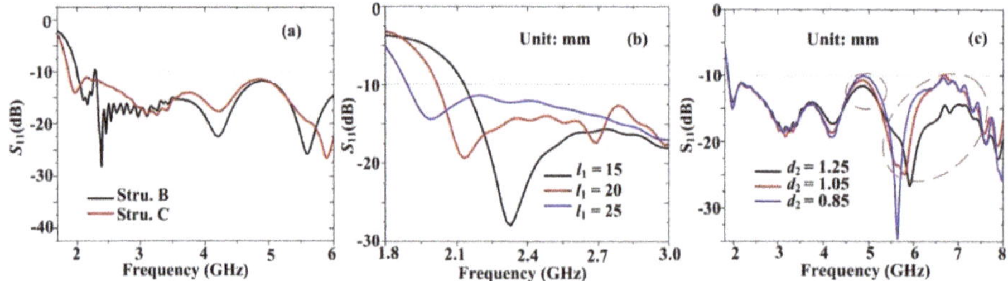

Figure 4. (**a**) Simulated reflection coefficients of Structure B and Structure C, (**b**) S_{11} for different maximum groove depths l_1 of Structure C, and (**c**) S_{11} for different groove gaps d_2 of Structure C.

It is expected that the etched grooves would change the surface currents and then the effective impedance. In order to understand the operating principle of the AVA and plasmonic AVA, surface current distributions at 1.8 GHz, 4 GHz, and 6 GHz are illustrated in Figure 5. First, it can be seen that the surface currents distribution at the region A of the AVA is larger than that of the plasmonic AVA at 1.8 GHz. After loading the periodic slots, electromagnetic waves are concentrated at the end of the coupling section and the inner end of the tapered slots (region B), as shown in Figure 5a. Hence, radiated electromagnetic energy can be concentrated in the axial direction of the tapered slot (*y*-direction). The same thing is true for the cases of the plasmonic AVA at 4 GHz and 6 GHz, as shown in Figure 5b,c.

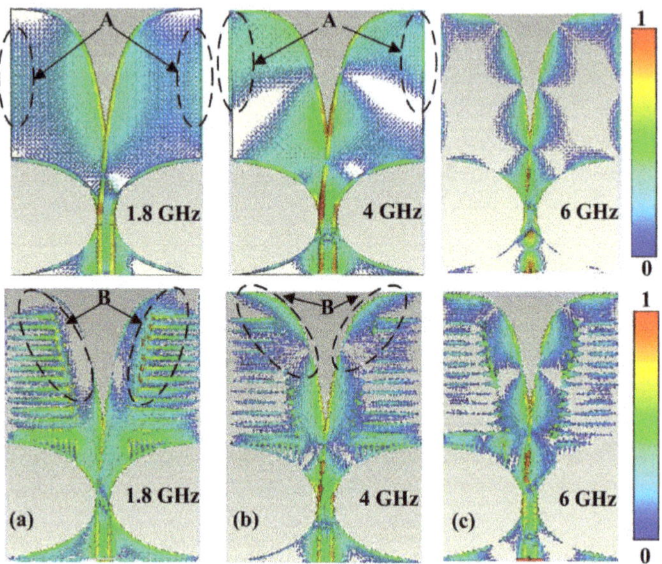

Figure 5. Surface current distributions of Structure C (**a**) at 1.8 GHz, (**b**) 4 GHz, and (**c**) at 6 GHz.

4. Results

In order to validate the proposed antenna, the plasmonic AVA has been fabricated and tested. A 50-Ω SMA connector was used to feed the antenna. The fabricated plasmonic AVA is shown in Figure 6a,b. The antennas were measured by using vector network analyzer (Keysight 8722ES), and the anechoic chamber operates from 600 MHz to 40 GHz.

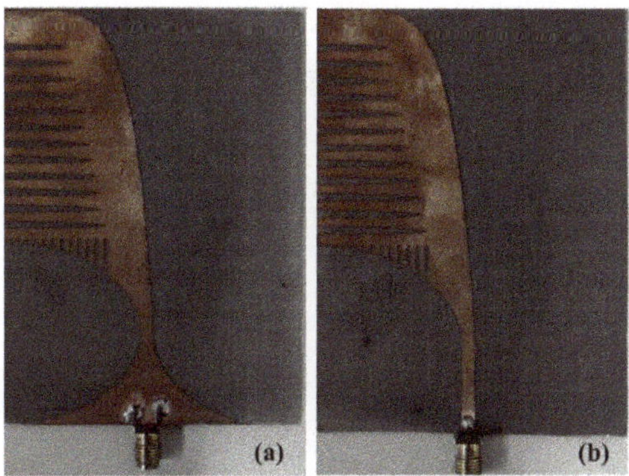

Figure 6. (**a**) The top view and (**b**) the bottom view of the fabricated plasmonic AVA.

The simulated and measured reflection coefficients are illustrated in Figure 7a. It can be observed that the lowest operating frequency ($S_{11} \leq -10$ dB) of Structure A (typical AVA) is 2.4 GHz, while it is 1.8 GHz for Structure C (plasmonic AVA). The corresponding electrical sizes are $0.75\lambda_0$ and $0.56\lambda_0$, respectively, where λ_0 is the wavelength in the air corresponding to the lowest operating frequency. Hence, it can be seen that the electrical size of the plasmonic AVA has been reduced by 25.3 percent, compared with that of the typical AVA. The measured S_{11} of the plasmonic AVA is also plotted in Figure 7a. It can be observed that the measured results agree well with the simulation results. The simulated and measured gains of the AVA and the plasmonic AVA are shown in Figure 7b. Compared to the gains of the AVA, it can be seen that the gain of the plasmonic AVA has been increased at the frequencies lower than 5.5 GHz and decreased at frequencies higher than 5.5 GHz. The measurement results agree well with the simulated results for the plasmonic AVA. The gain is larger than 5.5 dBi over the operating frequency from 1.8 GHz to 6 GHz. The measured peak gain is 9.12 dBi at 5.4 GHz. The measured average gain is 7.24 dBi over the operating frequency band, which covers 2G/3G/4G/5G bands.

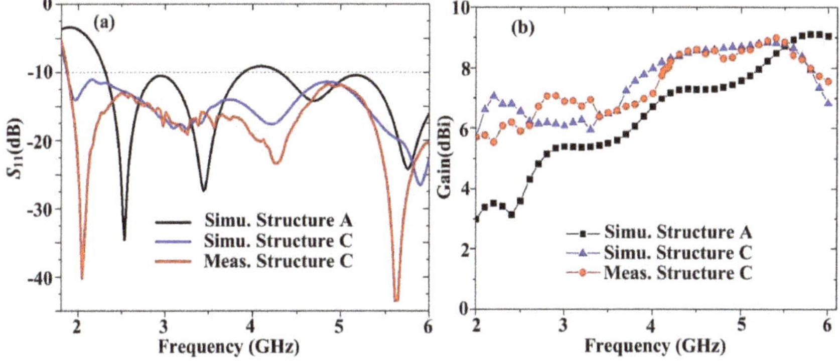

Figure 7. (**a**) The simulated and measured reflection coefficients of structures A and C (the typical AVA and the plasmonic AVA); (**b**) the simulated and measured gains of structures A and C.

The simulated and measured radiation patterns of the plasmonic AVA at 1.8 GHz, 4 GHz, and 6 GHz are illustrated in Figure 8. The upper and lower panels are the E-plane and H-plane radiation patterns, respectively. The half-power beam width (HPBW) in E-plane of the AVA and plasmonic AVA is shown in the Table 1. It can be seen that the radiation patterns of the plasmonic AVA are stable, compared with the results of the typical AVA.

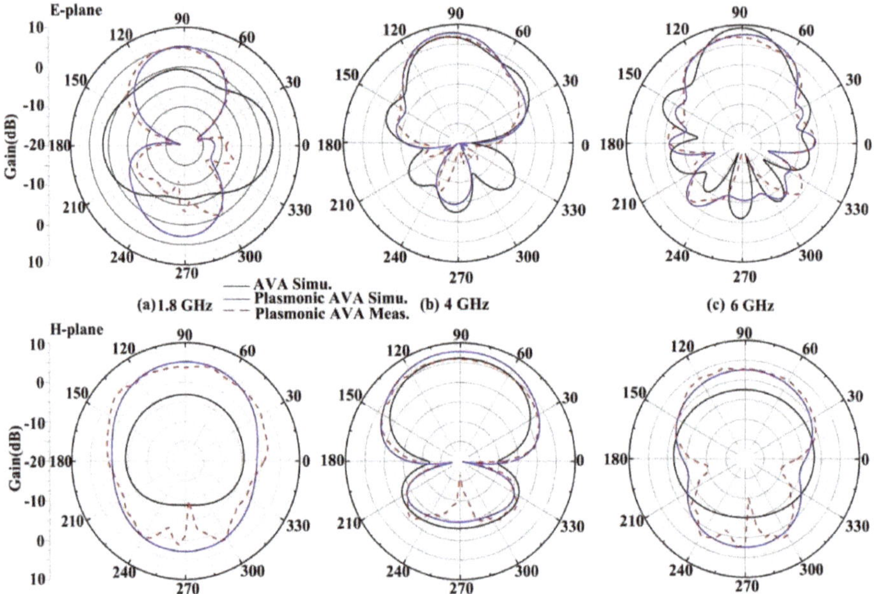

Figure 8. Simulated and Measured radiation patterns of the plasmonic AVA at (**a**) 1.8 GHz, (**b**) 4 GHz, and (**c**) 6 GHz.

Table 1. Simulated half-power bandwidth (HPBW) of the AVA and plasmonic AVA.

Frequency	Typical AVA	Plasmonic AVA
1.8 GHz	113.5	52.3
2.0 GHz	66.6	54.7
3.0 GHz	61	52.9
4.0 GHz	52.2	52.6
5.0 GHz	48.1	47.1
6.0 GHz	41	66.9

5. Discussion

The quantitive comparison of similar antennas has been given in Table 2. For the 2G/3G/4G/5G base station applications, the wide bandwidth is necessary to cover the 2G, 3G, and 4G bands and the new sub-6 GHz 5G frequency bands. Hence, only sub-6 GHz frequency band of the proposed antenna is shown here. Besides, the gain, HPBW, and the electrical size should be considered simultaneously for the base station antenna. From Table 2, we can see that the lowest operating frequency of the proposed plasmonic AVA is the lowest (1.8 GHz). The electrical size is $0.56\lambda_0$, which is a little larger than that of Refs. [31–33], where λ_0 is the wavelength in the air corresponding to the lowest operating frequency. The minimum gain (5.5 dBi) is the highest, since high gain is necessary for the base station antenna.

Table 2. Comparison of the similar antennas.

Reference (Year)	Physical Size (mm^2)	f_{min} (GHz)	Gain (dBi)	Electrical Size
[31] (2018)	71×50	2.0	4–7.5	$0.47\lambda_0 \times 0.3\lambda_0$
[32] (2018)	50×40	2.8	5.5–9	$0.46\lambda_0 \times 0.37\lambda_0$
[33] (2011)	60×48	2.4	3.8–10	$0.48\lambda_0 \times 0.38\lambda_0$
[34] (2019)	90×80	3.76	5–7	$1.13\lambda_0 \times 1.0\lambda_0$
[35] (2019)	186×77	2.5	4–16	$1.55\lambda_0 \times 0.64\lambda_0$
[36] (2018)	60.7×57.5	3.3	3.8–12.6	$0.67\lambda_0 \times 0.63\lambda_0$
[37] (2017)	104×100	2.0	2.2–8	$0.69\lambda_0 \times 0.67\lambda_0$
Structure A	94×70	2.4	3–9	$0.75\lambda_0 \times 0.56\lambda_0$
Structure B	94×70	2.1	4.5–8	$0.66\lambda_0 \times 0.49\lambda_0$
Structure C	94×70	1.8	5.5–9	$0.56\lambda_0 \times 0.42\lambda_0$

6. Conclusions

Here, a broadband and miniaturized plasmonic antipodal Vivaldi antenna is designed and measured, where plasmonic metamaterials are loaded on the antipodal Vivaldi antenna to improve the antenna performance. Compared of the typical AVA, the electrical size of the plasmonic AVA is reduced. The operating frequency is from 1.8 GHz to 6 GHz, which covers the 2G/3G/4G/5G communication band. The simulated and measured results of the plasmonic agree well. The measured results show that the radiation pattern of the plasmonic AVA is stable and the average gain of the antenna is 7.24 dBi.

Author Contributions: Conceptualization, Y.J.Z.; data curation, L.H.D.; formal analysis, L.H.D.; methodology, L.H.D.; resources, C.T.; writing—original draft preparation, L.H.D.; writing—review and editing, Y.J.Z.; supervision, Y.J.Z.; project administration, Y.J.Z.; funding acquisition, C.T. and Y.J.Z. All authors have read and agreed to the published version of the manuscript.

Funding: This research was funded by the National Natural Science Foundation of China, grant number 61971469, and by Science and Technology Commission Shanghai Municipality (STCSM), grant numbers 18ZR1413500 and SKLSFO2017-05.

Conflicts of Interest: The authors declare no conflict of interest.

References

1. Li, C.; Luo, Q.; Mao, G.; Sheng, M.; Li, J. Vehicle-mounted base station for connected and autonomous vehicles: Opportunities and challenges. *IEEE Wirel. Commun.* **2019**, *3*, 31–37. [CrossRef]
2. Imran, M.A.; Sambo, Y.A.; Abbasi, Q.H. 5G communication systems and connected healthcare. In *Enabling 5G Communication Systems to Support Vertical Industries*; John Wiley & Sons: Hoboken, NJ, USA, 2019; pp. 149–177.
3. Yan, J.; Wu, D.; Wang, H.; Wang, R. Multipoint. Cooperative transmission for virtual reality in 5G new radio. *IEEE Multimed.* **2019**, *26*, 51–58. [CrossRef]
4. Wu, Q.; Liang, P.; Chen, X. A broadband ±45° dual-polarized multiple-input multiple-output antenna for 5G base stations with extra decoupling elements. *J. Commun. Inf. Netw.* **2018**, *3*, 31–37. [CrossRef]
5. Zheng, D.; Chu, Q. A wideband dual-polarized antenna with two independently controllable resonant modes and its array for base-station applications. *IEEE Antennas Wirel. Propag. Lett.* **2017**, *16*, 2014–2017. [CrossRef]
6. Wu, B.Q.; Luk, K.M. A broadband dual-polarized magneto-electric dipole antenna with simple feeds. *IEEE Antennas Wirel. Propag. Lett.* **2009**, *8*, 60–63.
7. Gou, Y.; Yang, S.; Li, J.; Nie, Z. A compact dual-polarized printed dipole antenna with high isolation for wideband base station applications. *IEEE Trans. Antennas Propag.* **2014**, *62*, 4392–4395. [CrossRef]
8. Huang, H.; Liu, Y.; Gong, S. A broadband dual-polarized base station antenna with sturdy construction. *IEEE Antennas Wirel. Propag. Lett.* **2017**, *16*, 665–668. [CrossRef]
9. Chu, Q.X.; Wen, D.L.; Luo, Y. A broadband ±45° dual-polarized antenna with Y-shaped feeding lines. *IEEE Trans. Antennas Propag.* **2015**, *63*, 483–490. [CrossRef]
10. Zheng, D.Z.; Chu, Q.X. A multimode wideband ±45° dual-polarized antenna with embedded loops. *IEEE Antennas Wirel. Propag. Lett.* **2017**, *16*, 633–636. [CrossRef]
11. Gao, S.; Li, L.W.; Leong, M.S.; Yeo, T.S. A broad-band dual-polarized microstrip patch antenna with aperture coupling. *IEEE Trans. Antennas Propag.* **2003**, *51*, 898–900. [CrossRef]

12. Sim, C.; Chang, C.; Row, J. Dual-feed dual-polarized patch antenna with low cross polarization and high isolation. *IEEE Trans. Antennas Propag.* **2009**, *57*, 3321–3324. [CrossRef]

13. Khan, M.; Chatterjee, D. Characteristic mode analysis of a class of empirical design techniques for probe-fed, U-slot microstrip patch antennas. *IEEE Trans. Antennas. Propag.* **2016**, *64*, 2758–2770. [CrossRef]

14. Zhang, Y.; Li, D.; Liu, K.; Fan, Y. Ultra-wideband dual-polarized antenna with three resonant modes for 2G/3G/4G/5G communication systems. *IEEE Access* **2019**, *7*, 43214–43221. [CrossRef]

15. Yu, Z.; Yu, J.; Ran, X.; Zhu, C. A novel Koch and Sierpinski combined fractal antenna for 2G/3G/4G/5G/WLAN/navigation applications. *Microw. Opt. Technol. Lett.* **2017**, *59*, 2147–2155. [CrossRef]

16. Wen, L.H.; Gao, S.; Mao, C.X.; Luo, Q.; Hu, W.; Yin, Y.; Yang, X. A wideband dual-polarized antenna using shorted dipoles. *IEEE Access* **2018**, *6*, 39725–39733. [CrossRef]

17. Van Rooyen, M.; Odendaal, J.W.; Joubert, J. High-gain directional antenna for WLAN and WiMAX applications. *IEEE Antennas Wirel. Propag. Lett.* **2017**, *16*, 286–289. [CrossRef]

18. Gibson, P.J. The vivaldi aerial. In Proceedings of the 9th European Microwave Conference, Brighton, UK, 17–20 September 1979; pp. 101–105.

19. Simons, R.N.; Dib, N.I.; Lee, R.Q.; Katehi, L.P.B. Integrated uniplanar transition for linearly tapered slot antenna. *IEEE Trans. Antennas Propag.* **1995**, *43*, 998–1002. [CrossRef]

20. Chen, Y.J.; Hong, W.; Wu, K. Design of a monopulse antenna using a dual V-type linearly tapered slot antenn. *IEEE Trans. Antennas Propag.* **2008**, *56*, 2903–2909. [CrossRef]

21. Ozbay, Z. Plasmonics: Merging photonics and electronics at nanoscale dimensions. *Science* **2006**, *311*, 189–193. [CrossRef]

22. Hibbins, A.P. Experimental verification of designer surface plasmons. *Science* **2005**, *308*, 670–672. [CrossRef]

23. Pendry, J.B.; Martin-Moreno, L.; Garcia-Vidal, F.J. Mimicking surface plasmons with structured surfaces. *Science* **2004**, *305*, 847–848. [CrossRef]

24. Fong, B.H.; Colburn, J.S.; Ottusch, J.J.; Visher, J.L.; Sievenpiper, D.F. Scalar and tensor holographic artificial impedance surfaces. *IEEE Trans. Antennas Propag.* **2010**, *58*, 3212–3221. [CrossRef]

25. Wang, X.; Li, Z.; Fei, X.; Wang, J. A holographic antenna based on spoof surface plasmon polaritons. *IEEE Antennas Wirel. Propag. Lett.* **2018**, *17*, 1528–1532. [CrossRef]

26. Dadgarpour, A.; Zarghoon, B.; Virdee, B.S.; Denidni, T.A. Improvement of gain and elevation tilt angle using metamaterial loading for millimeter-wave applications. *IEEE Antennas Wirel. Propag. Lett.* **2016**, *15*, 418–420. [CrossRef]

27. Shi, Y.; Li, K.; Wang, J.; Li, L.; Liang, C.H. An etched planar meta-surface half Maxwell fish-eye lens antenna. *IEEE Trans. Antennas Propag.* **2015**, *63*, 3742–3747. [CrossRef]

28. Hongnara, T.; Chaimool, S.; Akkaraekthalin, P.; Zhao, Y. Design of compact beam-steering antennas using a meta-surface formed by uniform square rings. *IEEE Access* **2018**, *6*, 9420–9429. [CrossRef]

29. Singh, A.K.; Abegaonkar, M.P.; Koul, S.K. High-gain and high-aperture-efficiency cavity resonator antenna using metamaterial superstrate. *IEEE Antennas Wirel. Propag. Lett.* **2017**, *16*, 2388–2391. [CrossRef]

30. Qin, F.; Zhang, Q.; Xiao, J. Sub-wavelength unidirectional antenna realized by stacked spoof localized surface plasmon resonators. *Sci. Rep.* **2016**, *6*, 29773. [CrossRef]

31. Huang, M.; Wang, L.; Qiao, W. Design of 2 to 18 GHz balanced antipodal Vivaldi antennas using substrate-integrated lenses. *Electromagnetics* **2018**, *38*, 478–487. [CrossRef]

32. Zhu, H.; Li, X.; Yao, L.; Xiao, J. A novel dielectric loaded Vivaldi antenna with improved radiation characteristics for UWB application. *Appl. Comput. Electromagn. Soc. J.* **2018**, *33*, 394–398.

33. Peng, F.; Yong, C.J.; Wei, H. A miniaturized antipodal Vivaldi antenna with improved radiation characteristics. *IEEE Antennas Wirel. Propag. Lett.* **2011**, *10*, 127–130. [CrossRef]

34. Marek, D.; Harihara, S.G.; Prabhu, S.S. Design and validation of an antipodal Vivaldi antenna with additional slots. *Int. J. Antennas Propag.* **2019**, *2019*, 7472186.

35. Eichenberger, J.; Yetisir, E.; Ghalichechian, N. High-Gain antipodal Vivaldi Antenna with pseudoelement and notched tapered slot operating at (2.5 to 57) GHz. *IEEE Trans. Antennas Propag.* **2019**, *67*, 4357–4366. [CrossRef]

36. Fayu, W.; Jun, C.; Binhong, L. A novel ultra-wideband antipodal Vivaldi antenna with trapezoidal dielectric substrate. *Microw. Opt. Technol. Lett.* **2018**, *60*, 449–455.
37. Yang, Z.; Jingjian, H.; Weiwei, W. An antipodal Vivaldi antenna with band-notched characteristics for ultra-wideband applications. *AEU Int. J. Electron. Commun.* **2017**, *76*, 152–157. [CrossRef]

Review

Planar Microwave Resonant Sensors: A Review and Recent Developments

Jonathan Muñoz-Enano [1,*], Paris Vélez [1], Marta Gil [2] and Ferran Martín [1]

[1] CIMITEC, Departament d'Enginyeria Electrònica, Universitat Autònoma de Barcelona, 08193 Bellaterra, Spain; paris.velez@uab.cat (P.V.); ferran.martin@uab.es (F.M.)

[2] Departamento de Ingeniería Audiovisual y Comunicaciones, Universidad Politécnica de Madrid, 28031 Madrid, Spain; Marta.Gil.Barba@upm.es

* Correspondence: Jonatan.Munoz@uab.cat

Received: 2 March 2020; Accepted: 7 April 2020; Published: 10 April 2020

Abstract: Microwave sensors based on electrically small planar resonant elements are reviewed in this paper. By virtue of the high sensitivity of such resonators to the properties of their surrounding medium, particularly the dielectric constant and the loss factor, these sensors are of special interest (although not exclusive) for dielectric characterization of solids and liquids, and for the measurement of material composition. Several sensing strategies are presented, with special emphasis on differential-mode sensors. The main advantages and limitations of such techniques are discussed, and several prototype examples are reported, mainly including sensors for measuring the dielectric properties of solids, and sensors based on microfluidics (useful for liquid characterization and liquid composition). The proposed sensors have high potential for application in real scenarios (including industrial processes and characterization of biosamples).

Keywords: microwave sensor; differential sensor; dielectric characterization; microfluidics; electrically small resonators; biosensors

---

1. Introduction

Within the paradigm of the internet of things (IoT), or more generally the internet of everything (IoE), and the advent of the fourth industrial revolution (also known as Industry 4.0), the use of sensors has experienced an exponential growth. Moreover, this trend will continue with the progressive implantation of the 5th generation of mobile networks, designated as 5G, which is expected to satisfy the large-scale connectivity requirements that today's modern society demands (an interesting example is the autonomous and connected vehicle, necessarily equipped with hundreds of sensors). There are many available technologies for the implementation of sensors. Aspects such as diverse as sensor cost, size, and complexity, as well as the type of measurand, among others, dictate the preferred option. Although the most extended technology in modern sensors is probably optics/photonics (examples include optical fiber sensors, laser-based sensors, wearable chemical and biological sensors, nanophotonic biosensors, image sensors, etc.), significant efforts have been dedicated in recent years to the research and development of microwave sensors, especially for applications related to material characterization and composition. Microwaves are very sensitive to the properties of the materials to which they interact. Therefore, microwave technology is very useful for material sensing. Moreover, microwaves exhibit further interesting properties for sensing, such as low-cost generation and detection systems, interaction to the materials at different scales (i.e., through the near-field or the far-field), compatibility with planar technology, wireless connectivity, and robustness against harsh environments, among others. Thus, highly sensitive, robust, low cost, small size, and low-profile microwave sensors and wireless sensors based on the above-cited properties can be implemented.

This review paper focuses on microwave sensors, and particularly on planar sensors based on resonant elements for material characterization, including solids or liquids. Planar sensors are of special interest as their low-profile is compatible with many applications, where bulk sensors (e.g., cavity sensors or waveguide-based sensors [1,2], among others) may find a severe limitation, e.g., conformal sensors [3], wearable sensors [4], submersible sensors [5], integrated sensors [6], lab-on-a-chip sensors [7], microfluidic sensors [8–11], etc. On the other hand, despite the fact that non-resonant planar sensors operating at microwave frequencies have been reported [12–14], the combination of sensor size and performance (sensitivity) of resonant-type sensor is difficult to achieve with non-resonant methods. Finally, to end this introductory section, let us mention that although probably most planar microwave resonant sensors have been devoted to material characterization (the main interest in this review paper, to be discussed in detail later), there are many reported realizations focused on motion control applications (linear and angular displacement and velocity measurements) [15–30].

2. Classification of Planar Microwave Resonant Sensors

The classification subject of this section obeys the operating principle of the considered sensors (alternatively, sensors can be categorized by their applications, or frequency range, etc.). Thus, according to their working principle, planar resonant sensors can be divided in four main types: (i) frequency variation sensors, (ii) coupling modulation sensors, (iii) frequency splitting sensors, and (iv) differential-mode sensors. Let us briefly describe their working principle in the next paragraphs.

Frequency variation sensors are based on the variation of the resonance frequency (and eventually quality factor or peak/notch magnitude) of a resonant element, caused by the measurand. The typical, although not exclusive, configuration of such sensors is a transmission line loaded with the resonant element (either in contact or coupled to it), see Figure 1a. Although examples of frequency variation sensors devoted to the measurement of spatial variables have been reported [15], typically these sensors have been applied to dielectric characterization of solids and liquids [31–46], as far as the resonance frequency and quality factor depend on the complex dielectric constant of the surrounding material (the so-called material under test (MUT)). These sensors are very simple, but they are potentially subjected to cross-sensitivities caused by variations in ambient factors, such as temperature and humidity, and therefore they need calibration before their use.

To alleviate or minimize the impact of cross-sensitivities, symmetry-based sensors are the solution [47]. As symmetry is invariant under potential changes in environmental factors, at least at the typical scales of the sensors, it is expected that sensors based on symmetry properties exhibit certain robustness against environment-related cross-sensitivities. Coupling modulation sensors belong to this category [16–27,48]. Such sensors are implemented by symmetrically loading a transmission line by means of a symmetric resonator (Figure 1b). However, the combination of line and resonant element should not be arbitrary. Namely, it is necessary that the symmetry plane of the resonator and the one of the line are of different electromagnetic sort, i.e., one must be an electric wall and the other one must be a magnetic wall. By this means, if symmetry is preserved, line-to-resonator coupling is prevented [47], the resonator is not excited, and the line is transparent, i.e., the frequency response does not exhibit any notch (lack of resonance). However, by truncating symmetry (e.g., by means of an asymmetric dielectric loading or through a relative linear or angular displacement between the line and the resonant element), line-to-resonator coupling arises and, consequently, the resonance appears. The magnitude of the notch (the typical output variable) is determined by the coupling degree, intimately related to the level of the asymmetry. Thus, coupling modulation by symmetry disruption is useful for sensing purposes. Nevertheless, as the most natural procedure for symmetry truncation is by a relative displacement between the resonator and the line, most sensors based on this principle have been devoted to angular and linear measurements [16–27,48]. For such application, the most limitative aspect of these sensors is the input dynamic range, related to the small dimensions of the sensing resonators. A variant of this approach, providing the solution to such limitation, was presented in [28–30], where circular chains of resonant elements were applied to the implementation of microwave rotary encoders based on pulse counting (linear encoders, potentially exhibiting an unlimited dynamic range, were also presented in [49,50]). Another drawback of these sensors is related to the fact that measuring the notch magnitude is more sensitive to the effects of noise, as compared to frequency measurements. However, a small frequency scan, or even a single-frequency measurement, suffices to collect the sensor information.

Frequency splitting sensors combine the advantages of frequency measurements (noise tolerant) and symmetry properties (robustness against cross-sensitivities) [24,51–59]. In these sensors, a symmetric transmission line-based structure is symmetrically loaded with a pair of (not necessarily symmetric) resonators (Figure 1c). If symmetry is preserved, a single notch in the transmission coefficient arises. However, if symmetry is truncated, e.g., by an asymmetric dielectric loading, the original notch splits in two notches separated a distance that depends on the level of asymmetry. Thus, in frequency splitting sensors, the output variable is such frequency separation, and, eventually, the difference in the notch magnitude (two variables are needed, for instance, for the measurement of the dielectric constant and loss tangent of materials [56,58]). A potential weakness of these sensors is the limited sensitivity and resolution if the resonant elements are coupled [51,52,55]. Nevertheless, solutions to this problem have been reported, as discussed in [56,57], at the expense of significantly separating the sensing resonators (either by using a splitter/combiner scheme [57,58] or a cascaded topology [56]). Although these sensors are not true differential sensors (in the sense that two independent sensors are not used), the input and the output variable are typically differential variables. The differential dielectric constant or loss tangent (between a reference (REF) sample, or material, and the material under test (MUT)) is the usual input variable, whereas the differential notch frequency and/or magnitude is the natural output variable/s. Moreover, frequency splitting sensors can be easily used as comparators, as they are able to detect differences between the REF and MUT samples. For this application, sensor resolution is the key aspect. However, in terms of resolution, such sensors cannot compete, in general, against true differential sensors, to be discussed next.

Differential-mode sensors also belong to the group of symmetry-based sensors. Such devices are composed of two independent sensors, one sensitive to the REF sample (or measurand in general), and the other one sensitive to the MUT sample (Figure 1d). These sensors can operate as single-ended sensors as well, e.g., based on frequency variation, where only one of the individual sensing elements is used. However, exploiting the differential input and output variable/s provides robustness against cross-sensitivities. As far as independent sensing elements are used, the abovementioned coupling effects, which typically limit the resolution, can be prevented in differential sensors. The output variable in these sensors is merely the difference between the output variable of each sensing element. It has been demonstrated that highly sensitive differential sensors for dielectric constant measurements can be implemented by means of a pair of meander lines. These sensors are based on the measurement of the phase difference between the pair of lines, and the penalty to achieve high sensitivity is sensor size [14]. Nevertheless, it has been demonstrated that by replacing the meander lines with artificial lines, sensitivity optimization can be made compatible with moderate or even small sizes [12,60]. Many differential sensors based on resonator-loaded lines (the interest in this work) have been reported, including microfluidic sensors for liquid characterization [61–67]. In some realizations, the output variable is the cross-mode transmission coefficient [62–67], proportional to the difference between the transmission coefficients of both sensing lines, provided such lines are uncoupled. Huge sensitivities and resolutions, useful to detect, e.g., small volume fractions or electrolyte concentrations in liquid solutions, have been achieved by means of these sensors. Although, typically, differential-mode sensors involve four-port measurements (these sensors are four-port devices), it has been demonstrated that by adding extra (microwave) circuitry, differential sensing by means of two-port devices (and measurements) can be achieved [68,69]. In [69], the differential sensor works in reflection (other reflective-mode sensors have been reported [70], also including single-ended sensors [71]). The sensor reported in [72] is also remarkable, where symmetry truncation in a pair of slotted resonators generates quasi-microstrip to slot-mode conversion, useful for sensing.

Although examples of the different sensor types highlighted in this section will be provided, the main emphasis in this paper is on differential-mode sensors. Nevertheless, before reviewing the different implementations, let us first present the potential candidates for resonant sensing elements, and discuss their advantages and disadvantages.

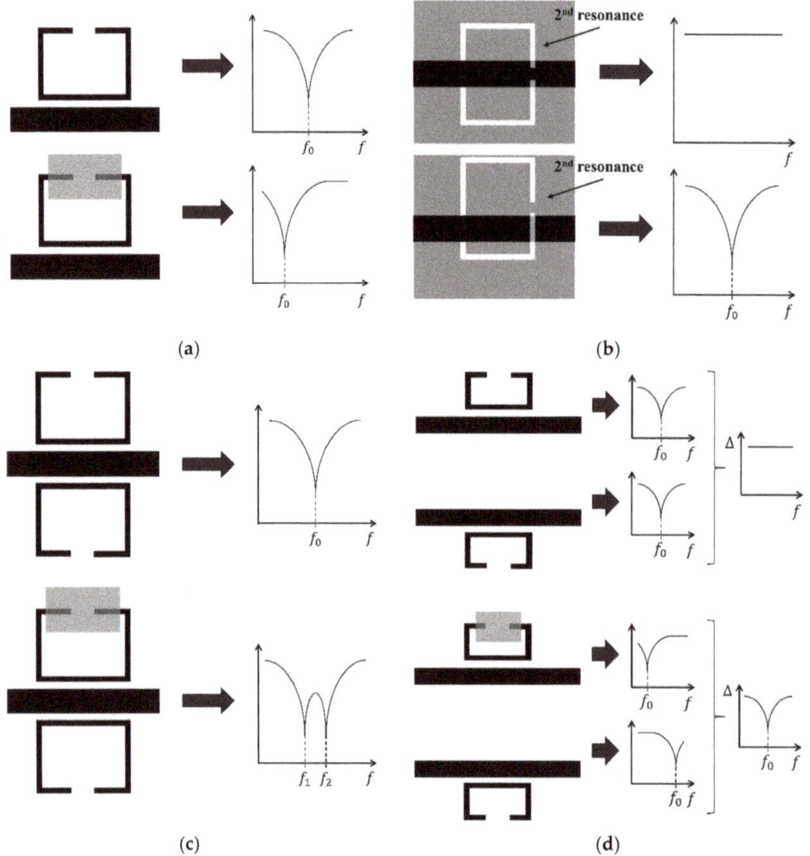

Figure 1. Typical topology and working principle of the considered planar microwave resonant sensors. (**a**) Frequency variation; (**b**) coupling modulation; (**c**) frequency splitting; (**d**) differential-mode.

3. Planar Resonant Elements for Sensing

There are dozens of planar resonant elements that can be useful for sensing purposes. In this paper, the interest is in electrically small (quasi-lumped) resonators that can be described by means of a lumped-element equivalent circuit model up to frequencies beyond their fundamental resonance (the authors recommend the books [73,74], and the paper [75], where a review of many of these resonators is carried out). Probably the simplest classification of such resonators can be made by differentiating on whether they are metallic or slot resonators. In the first group, resonant elements of interest for sensing include the step impedance resonator (SIR) [76], the step impedance shunt stub (SISS) [77], the split-ring resonator (SRR) [78], the open split-ring resonator (OSRR) [79], and the electric-LC (ELC) resonator [80], among others (see Figure 2). Slot resonators are typically etched in the ground plane and are sometimes designated as defect ground structure (DGS) resonators. Examples of such resonators include the complementary counterparts of the SRR and the OSRR, that is, the complementary split-ring resonator (CSRR) [81] and the open complementary split-ring resonator (OCSRR) [82], respectively, the dumbbell-shaped DGS (DB-DGS) resonator [83], and the magnetic-LC (MLC) resonator [84], among others (see Figure 3). There are many other electrically small resonators, including the S-shaped-SRR [85], the spiral resonator [86], the non-bi-anisotropic-SRR [87] (as well as their complementary counterparts), the broadside-coupled-SRR [88], etc., but their use for sensing has been more limited.

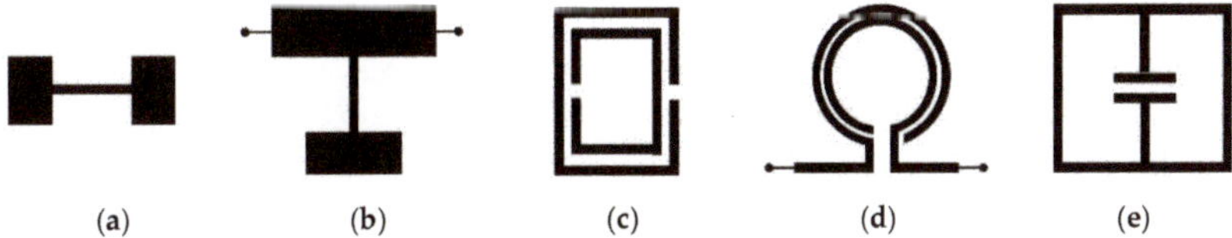

(a)   (b)   (c)   (d)   (e)

Figure 2. Typical topology of some electrically small metallic resonators useful for sensing. (**a**) Step impedance resonator (SIR), (**b**) step impedance shunt stub (SISS), (**c**) split-ring resonator (SRR), (**d**) open split-ring resonator (OSRR), (**e**) electric-LC (ELC).

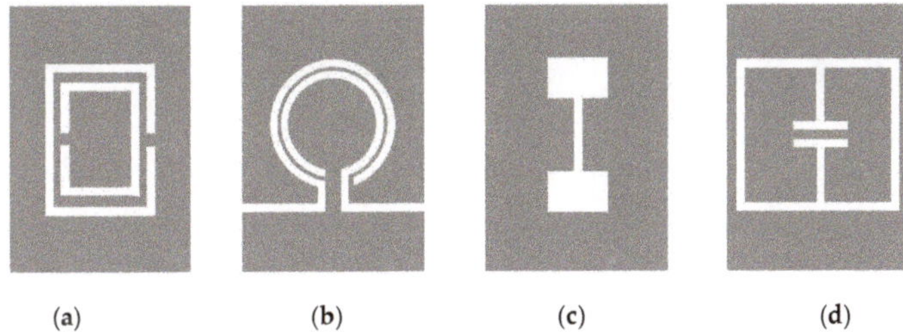

(a)   (b)   (c)   (d)

Figure 3. Typical topology of some electrically small slot resonators useful for sensing. (**a**) Complementary split-ring resonator (CSRR), (**b**) open complementary split-ring resonator (OCSRR), (**c**) dumbbell-shaped DGS (DB-DGS), (**d**) magnetic-LC (MLC).

A comparative analysis of electrically small resonators, with the focus on metamaterial and metamaterial-inspired circuit design, was carried out in [75]. For sensing, the most important aspect is the effect of resonator topology on sensor performance, mainly sensitivity, resolution, and dynamic range. Nevertheless, compatibility of the sensing element (resonator), including the dielectric load, with the required hardware and mechanical parts of the sensor is also important. For instance, for microfluidic applications, the fluidic channels plus the required mechanical accessories should not affect the transmission lines used for resonator's excitation. In this regard, slot resonators constitute, in general, a good solution, as the ground plane provides backside isolation, thereby preventing for any potential effect of the dielectric load on sensor functionality. The size of the resonant element may be also important in certain applications devoted to the characterization of small dielectric loads. Obviously, increasing the frequency is a possibility, but this has the effect of raising the cost of the associated electronics. Therefore, considering resonant elements with small electrical size is expected to provide a good tradeoff, allowing for the implementation of compact sensors (sensitive element) operating at moderate (or even low) frequencies. In this regard, the OSRR and the OCSRR are electrically smaller than the SRR and the CSRR, respectively, by a factor of two [74,89], and are therefore good candidates from the point of view of sensor size.

Concerning sensitivity and resolution (intimately related), it is difficult to firmly establish the preferred resonator topology, as it may be determined by several aspects, including the sensing working principle, the considered input and output variables, etc. For instance, for complex permittivity measurements, both the real and the imaginary part of the complex permittivity (input variables) must be determined, and two output variables (e.g., the frequency position and notch magnitude) are needed to unequivocally infer both measurands. For this kind of application, with more than one input variable (or measurand), several sensitivities, including cross-sensitivities, can be defined, and this further prevents from obtaining concluding remarks with regard to the effects of resonator topology on sensitivity and resolution. However, the real part of the complex permittivity (or the dielectric constant) of the MUT mainly affects the resonance frequency of the sensing element (through the effects on the capacitance), whereas the imaginary part (related to the loss tangent) primarily affects the quality factor or the notch magnitude. For a given resonator, the strategy for optimizing such "natural sensitivities" (in terms of geometry) typically coincides (e.g., for the OCSRR and dumbbell-DGS, as it was demonstrated in [62,64], respectively).

It is difficult, however, to determine absolutely the optimum resonator (among those depicted in Figure 2) in terms of sensitivity (and resolution) optimization. Nonetheless, to the light of a comparative analysis of the circuit model of the considered resonators, it is possible to infer some helpful hints. As an example, let us compare three slot resonators: the CSRR, the OCSRR, and the DB-DGS. The circuit models of such resonant elements loading a transmission line are depicted in Figure 4.

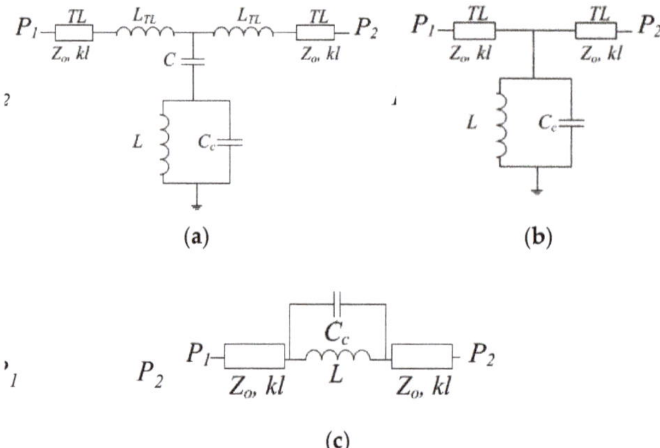

Figure 4. Circuit models of the CSRR- (**a**), OCSRR- (**b**), and DB-DGS-loaded line (**c**).

Let us assume that the resonant element (the sensing part) is loaded with a dielectric material slab (MUT) of sufficient thickness to ensure that the field lines generated in the slot do not reach the opposite interface, and let us consider that the substrate thickness satisfies also this requirement. If these sensors are intended to work based on frequency variation, the key figure of merit is the relative sensitivity of the resonance frequency, f_0, with the dielectric constant of the MUT, ε_{MUT}, given by

$$S = \frac{1}{f_0}\frac{df_0}{d\varepsilon_{MUT}} \tag{1}$$

where the resonance frequency is given by

$$f_0 = \frac{1}{2\pi}\left\{L\left(C + C_c\frac{\varepsilon_r + \varepsilon_{MUT}}{\varepsilon_r + 1}\right)\right\}^{-1/2} \tag{2}$$

Expression (2) is valid for the CSRR-loaded line, whereas for the OCSRR- and the DB-DGS-loaded line, the previous expression is valid by skipping C, the coupling capacitance between the line, and the CSRR. The unloaded resonators are described by the capacitance C_c and by the inductance L. Notice that if the dielectric constant of the MUT is different than the one of air (i.e., $\varepsilon_{air} = 1$), the result is a variation of the capacitance of the resonator (and hence a shift in f_0), as expected. In (1), ε_r is the dielectric constant of the substrate. Introducing (2) in (1), the relative sensitivity is found to be

$$S = -\frac{1}{2}\frac{C_c}{C_c(\varepsilon_r + \varepsilon_{MUT}) + C(\varepsilon_r + 1)} \tag{3}$$

In view of (3), we can conclude that it is convenient to reduce C and ε_r as much as possible for sensitivity optimization. Consequently, the DB-DGS and the OCSRR are good candidates for the implementation of high sensitive frequency variation sensors, at least as compared to CSRRs. In [90], it was demonstrated that the relative sensitivity of DB-DGS-loaded lines is better than the one of CSRR-loaded lines. To gain further insight on this, we have designed an OCSRR-loaded line (Figure 5a), and we have inferred the relative sensitivity as defined in (1), from the simulated responses obtained by considering MUTs of different dielectric constant. The results, depicted in Figure 6, and compared to those reported in [90], confirm that the relative sensitivity of the OCSRR-loaded line is comparable to the one of the DB-DGS, as predicted by the theory. Note that one difference between the OCSRR- and the CSRR- or the DB-DGS-loaded line is the type of response, with a maximum at f_0 for the OCSRR-based line, and notched at resonance for the other lines.

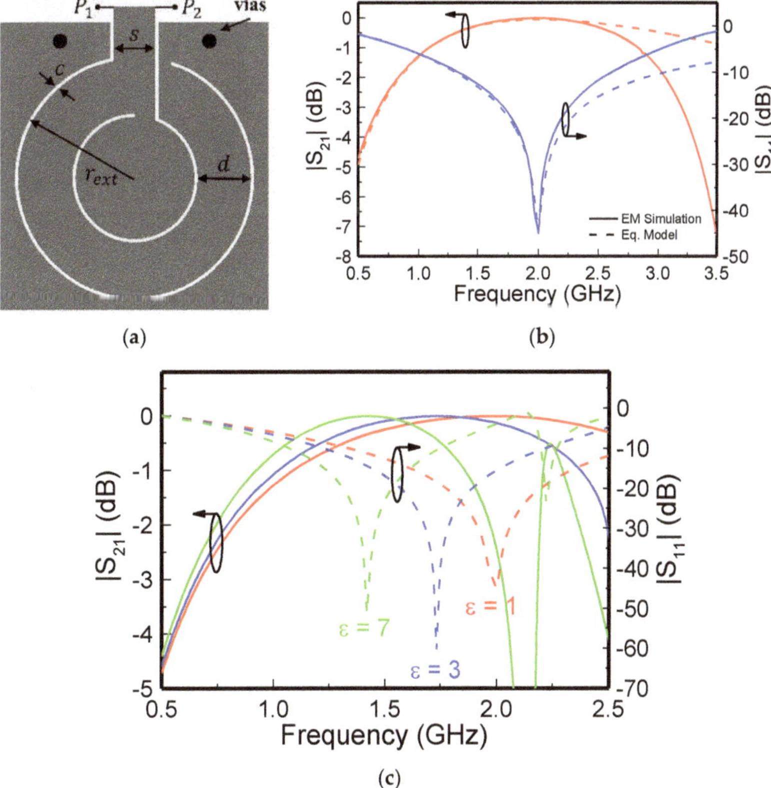

Figure 5. Topology of the designed OCSRR-loaded microstrip line (**a**), frequency response of the structure inferred from circuit and electromagnetic simulation (**b**), and simulated frequency response for various MUTs with the indicated dielectric constants (**c**). Dimensions (in mm) are $r_{ext} = 4$, $c = 0.1$, $d = 1.8$ and $s = 1.4$. The considered substrate is the *Rogers RO4003C* with dielectric constant $\varepsilon_r = 3.55$ and thickness $h = 1.524$ mm. Circuit parameters are $C_c = 1.23$ pF and $L = 5.1484$ nH. The ports in panel (**a**) can be separated (if needed for connector soldering) by merely adding 50 Ω access lines.

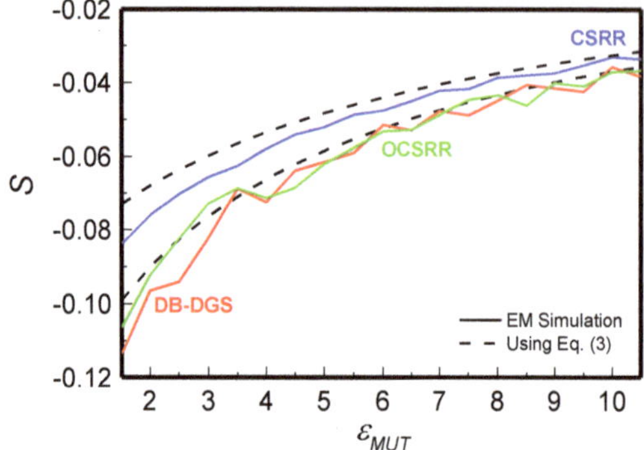

Figure 6. Relative sensitivity of the OCSRR-loaded line, as compared to those reported in [90].

4. Prototype Examples

In this section, several examples of planar microwave resonant sensors, categorized according to the classification of Section 2, are reported. These sensors are devoted to dielectric characterization, with the exception of the coupling modulation sensor of Section 4.2, a rotary encoder.

4.1. Frequency Variation Sensors

The first example is a frequency variation sensor equipped with a microfluidic channel for liquid characterization [46]. In this sensor, the sensitive element is a SISS resonator, loading a microstrip line. The most relevant aspect of this sensor is the high achieved relative sensitivity, by virtue of the specific arrangement, where the ground plane beneath the SISS is removed, and a metallization is added on the top surface, surrounding the rectangular SISS patch with a gap of width s (see Figure 7). Such metallization is connected to the backside ground plane by means of metallic vias. By these means, the single capacitance of the resonant element is the edge capacitance of the SISS, which can be easily perturbed by means of the MUT (either a solid or a liquid), placed on top of it. Note that with this configuration, any extra capacitance that degrades the sensitivity is removed (see Section 3). This sensor has been equipped with a microfluidic channel, in order to perform measurements of the complex permittivity of liquids (Figure 8).

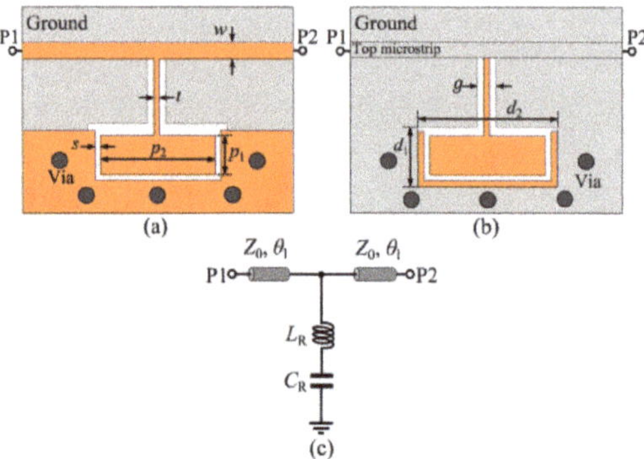

Figure 7. Topology of the SISS-based frequency variation sensor reported in [46], including the top (**a**) and bottom (**b**) views, and lumped element equivalent circuit model (**c**). From [46], reprinted with permission.

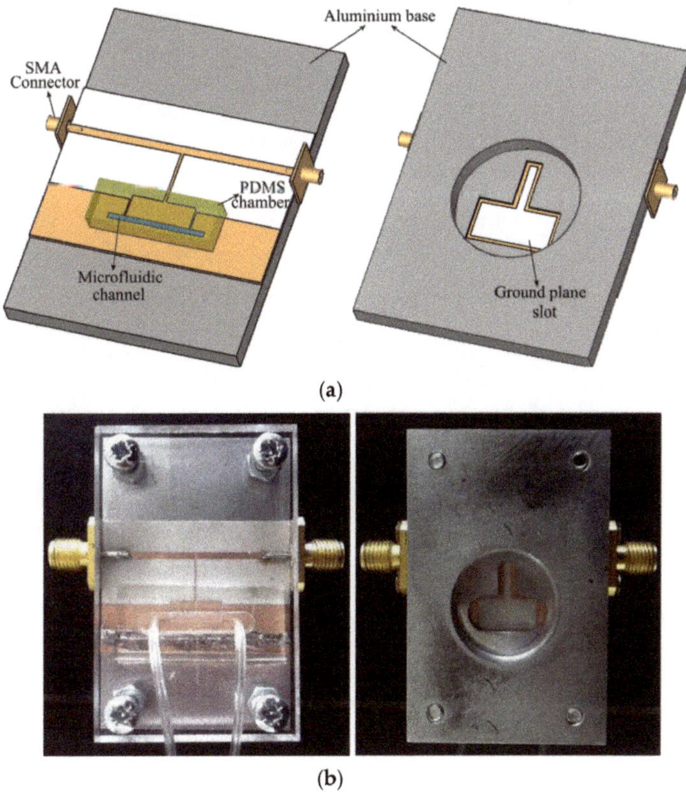

Figure 8. Schematic views (**a**) and photographs (**b**) of the fabricated sensor, including the top (**left**) and bottom (**right**) views. From [46], reprinted with permission.

The measured frequency responses of the sensor for different mixtures of DI water and ethanol are depicted in Figure 9a, whereas Figure 9b shows the resonance frequency position and the magnitude of the notch as a function of the volume fraction of water. The water–ethanol solutions offer a wide range of complex permittivity values, which are appropriate for sensor calibration. That is, with this set of measurements, a mathematical model for the sensor can be developed. Specifically, a mathematical relation linking the frequency shift and the notch magnitude to the complex relative permittivity of the test water–ethanol solutions was derived in [46]. As the complex permittivities of water–ethanol as a function of the volume fraction can be independently inferred, a nonlinear least square curve fitting in MATLAB can be used in order to derive an equation describing the relation between variations of the resonance frequency and notch magnitude as a function of the complex permittivity variations.

In [46], the authors used the complex permittivity values of water-ethanol mixtures provided in [91]. With such model, given in [46], the complex permittivity of other liquid solutions can be inferred from the measured values of the resonance frequency and notch magnitude. For instance, in [46], the authors characterized solutions of water–methanol. The measured resonance frequency and notch magnitude for the different concentrations is shown in Figure 10, whereas Figure 11 depicts the real and the imaginary part of the permittivity inferred from the mathematical model. In this figure, the values inferred from the literature are also depicted, and the good agreement between both sets of data can be appreciated. It is remarkable that the relative sensitivity of this sensor is very good, as compared to the one of similar sensors reported in the literature (see [46] for further details).

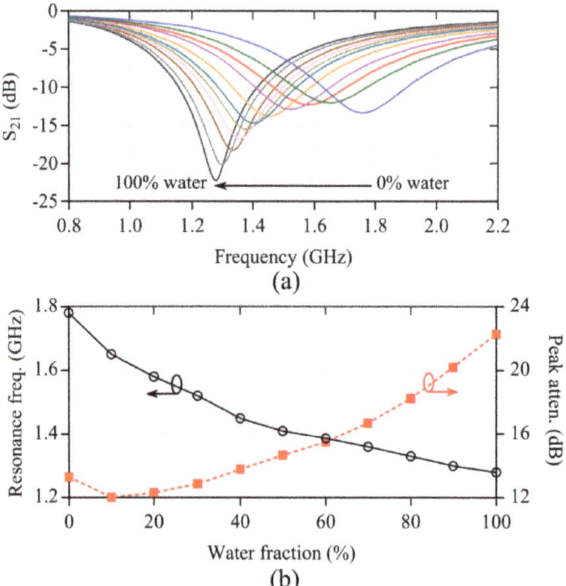

Figure 9. Measured frequency response of the sensor for various water–ethanol solutions (**a**) and representation of the resonance frequency and notch magnitude as a function of the water concentration (**b**). From [46], reprinted with permission. In panel (a), the step variation of water content is 10%.

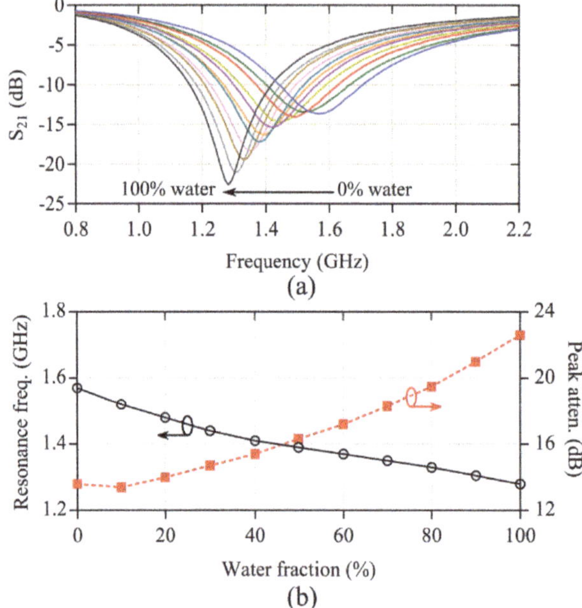

Figure 10. Measured frequency response of the sensor for various water-methanol solutions (**a**) and representation of the resonance frequency and notch magnitude as a function of the water concentration (**b**). From [46], reprinted with permission. In panel (a), the step variation of water content is 10%.

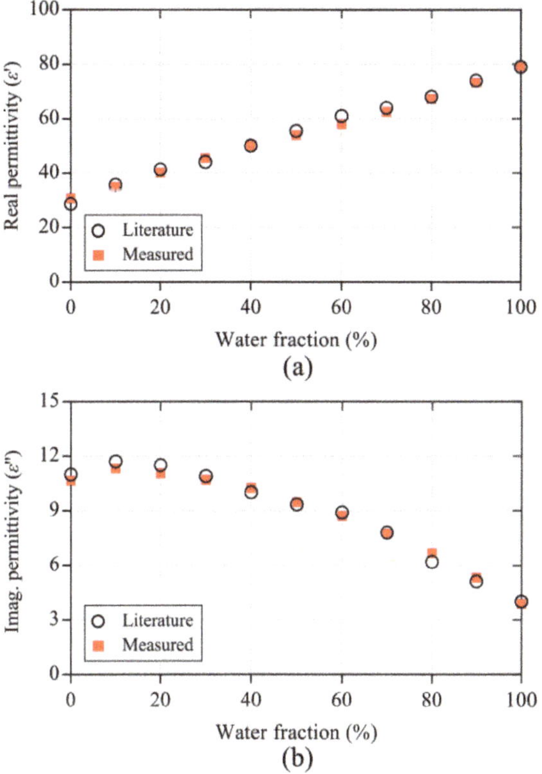

Figure 11. Real (**a**) and imaginary (**b**) part of the complex dielectric constant of water–methanol solutions as a function of the water content, inferred from the data of Figure 10 and the mathematical model developed in [46]. From [46], reprinted with permission.

4.2. Coupling Modulation Sensors

As discussed in Section 2, coupling modulation sensors are implemented by symmetrically loading a transmission line with a symmetric resonator. The symmetry planes of the line and resonant element should be of different electromagnetic sort, i.e., one should be a magnetic wall and the other one an electric wall. Figure 12 illustrates the working principle of coupling modulation sensors, by considering a CPW loaded with a SRR [47]. The line exhibits a magnetic wall at the axial symmetry plane for the fundamental CPW mode, whereas the symmetry plane of the SRR is an electric wall at the fundamental resonance. For perfect symmetry, the electromagnetic field generated by the line is not able to excite the SRR, and the structure is all-pass in the vicinity of the SRR fundamental resonance. This situation is preserved if the SRR is symmetrically loaded with a dielectric material. However, if symmetry is truncated, e.g., by means of an asymmetric dielectric load, as depicted in Figure 12, then the resonator is driven by the line (i.e., line-to-resonator coupling is activated), and a notched response (bandstop-type) appears. Although this type of sensor can be used for dielectric characterization or for comparison purposes by exploiting their electromagnetic symmetry properties, most coupling modulation sensors have been applied to the implementation of linear or angular displacement and velocity sensors [16–30,48]. The reason is that a relative linear or angular displacement between the line and the resonant element also truncates symmetry (except if the resonant element displaces linearly in the direction of the line axis). Thus, in this subsection, this sensing approach is illustrated by means of two examples, to be discussed next.

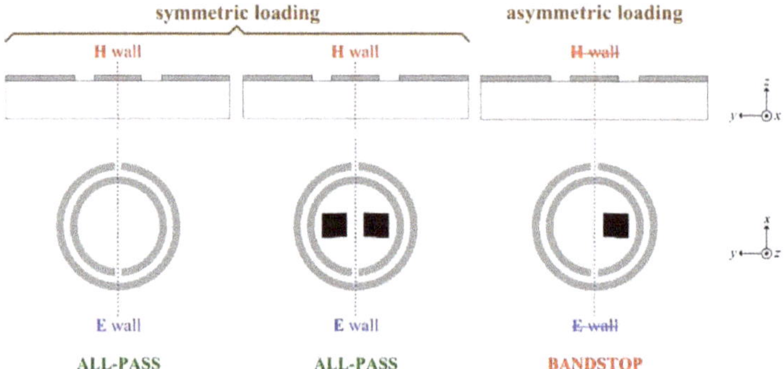

Figure 12. Illustration of symmetry disruption by means of square-shaped inclusions in a CPW loaded with a SRR. From [47], reprinted with permission.

In the first example [22], an angular displacement and velocity sensor based on a circular-shaped CPW loaded with a movable circular ELC resonator is reported. The ELC resonator is a bisymmetric particle, exhibiting orthogonal electric and magnetic walls. Thus, significant variation of line-to-resonator coupling by ELC rotation can be achieved, and this is interesting in terms of the output dynamic range. Figure 13 shows the picture of the CPW and ELC resonator, as well as the experimental set-up used for measuring the angular displacement and velocity.

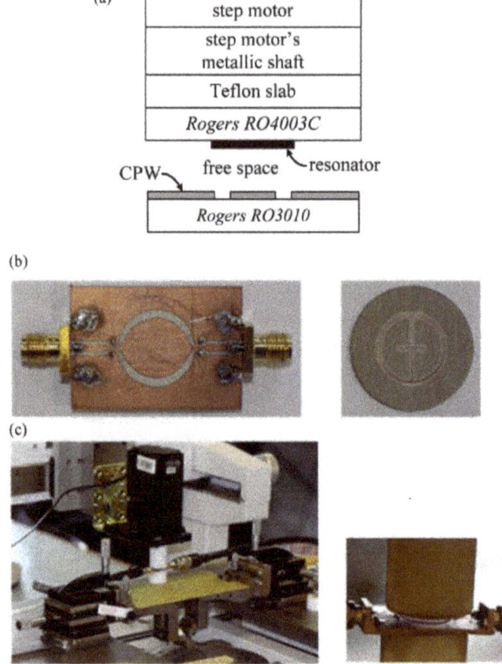

Figure 13. Set-up for the angular displacement and velocity measurement. (**a**) Cross section view, (**b**) photographs of the rotor (circular ELC resonator) and stator (CPW), and (**c**) photograph of the step-motor providing rotor motion and zoom view of the stator-rotor. From [22], reprinted with permission.

By rotor motion with regard to the stator, the transmission coefficient at f_0, the resonance frequency of the ELC, experiences a roughly linear variation with the rotation angle, as can be appreciated in Figure 14. Thus, the output variable for measuring angular displacement is the magnitude of the transmission coefficient. The sensitivity is reasonable, but the dynamic range is restricted to 90°, as it can be explained from obvious symmetry considerations. To measure the (average) angular velocity, the strategy proposed in [22] is based on an amplitude modulation (AM) scheme. By feeding the line with a harmonic signal tuned to the resonance frequency of the ELC, the amplitude at the output port is expected to be modulated due to the time-varying coupling between the line and the resonant element, and two pulses of the envelope function per cycle are expected. Thus, from the time distance between adjacent pulses, the angular velocity can be inferred. The envelope function can be simply obtained by means of an envelope detector, preceded by an isolator (implemented by means of a circulator) in order to protect the CPW line against mismatching reflections from the diode (a highly nonlinear device). The schematic diagram of the approach is presented in Figure 15. Figure 16 depicts a pair of recorded envelope functions (visualized in an oscilloscope in [22]), corresponding to different angular velocities of the rotor (particularly, 1 rpm and 50 rpm). The measured envelope functions reveal that the method is useful to provide the angular velocity. However, it is not possible to obtain instantaneous velocities with this approach, as only two pulses per cycle are generated. Namely, only the average angular speed corresponding to time lapses between adjacent pulses can be measured. In order to improve this limitative aspect, a very large number of pulses is necessary (like in optical encoders [92–94]). Thus, a different approach for the measurement of angular displacements and velocities was reported in [28], where a completely different configuration (edge configuration), providing hundreds of pulses per revolution, was implemented.

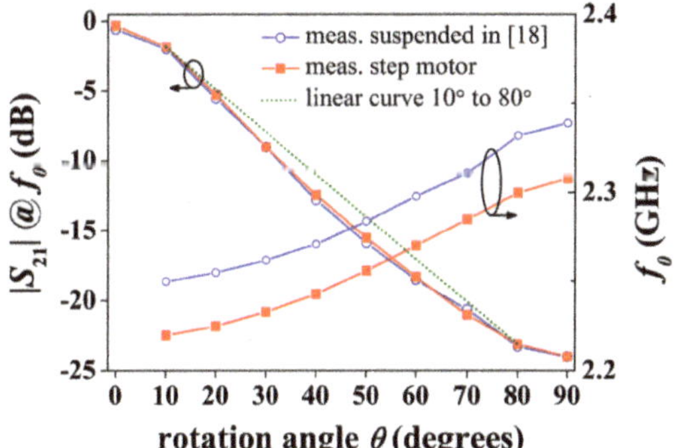

Figure 14. Transmission coefficient at f_0 as a function of the rotation angle. From [22], reprinted with permission.

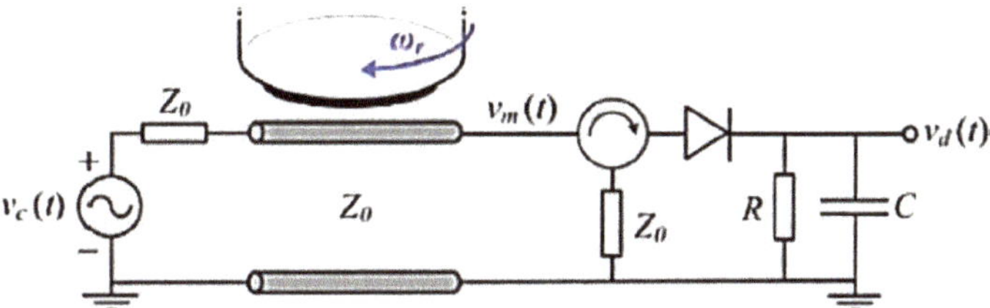

Figure 15. Sketch of the system for the measurement of angular velocities based on an AM modulation scheme.

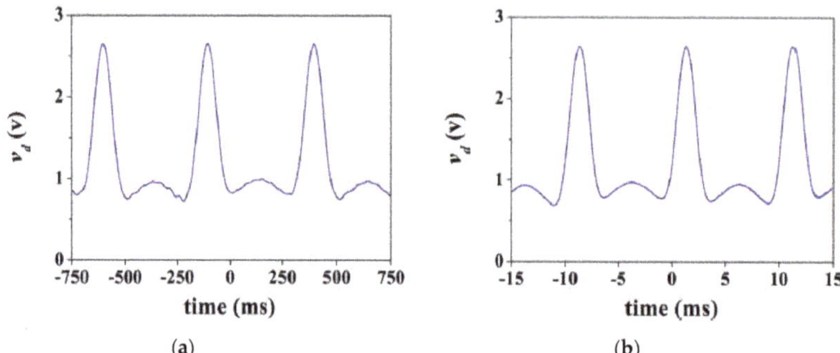

Figure 16. Measured envelope functions. (**a**) 1 rpm; (**b**) 50 rpm. From [22], reprinted with permission.

In the rotation sensor of Figure 13, the axis of the resonator coincides with the axis of the rotor (axial configuration). In [28], a rotor consisting of a circular chain of resonant elements etched along its edge was proposed as a means to boost the number of pulses (see Figure 17). The working principle is indeed very similar to the one used to infer the angular velocity with the sensor based on the axial configuration. Namely, by rotor motion, the amplitude of the feeding harmonic signal is modulated, but at least as many pulses as resonators in the chain can be generated per each revolution. Indeed, it was demonstrated in [29,30] that by conveniently choosing the frequency of the feeding signal, up to two pulses per resonant element can be achieved, and the sensor reported in [30] exhibits 1200 pulses per revolution (i.e., 600 resonators), which is a relevant figure of merit. Moreover, this sensor is equipped with a resonator's chain devoted to detect the motion direction (see details in [30]). Note that in this sensor based on the so-called edge configuration, the working principle is coupling modulation as well. By this means, the transmission coefficient of the line at the frequency of the feeding signal periodically varies with time, generating a pulsed AM signal at the output port of the line. However, coupling modulation is achieved by the consecutive crossings of the resonant elements of the chain over the axis of the line (stator), when the rotor is in motion. The working principle is very similar to the one of optical encoders, with apertures in a metallic disc. In such encoders, the apertures are detected by means of an optical beam (when an aperture lies in the path between the source and the detector, a pulse is generated). By contrast, in the sensor based on the edge configuration, the resonators are detected by means of an electromagnetic signal. Therefore, such sensors can be designated as electromagnetic (or microwave) rotary encoders.

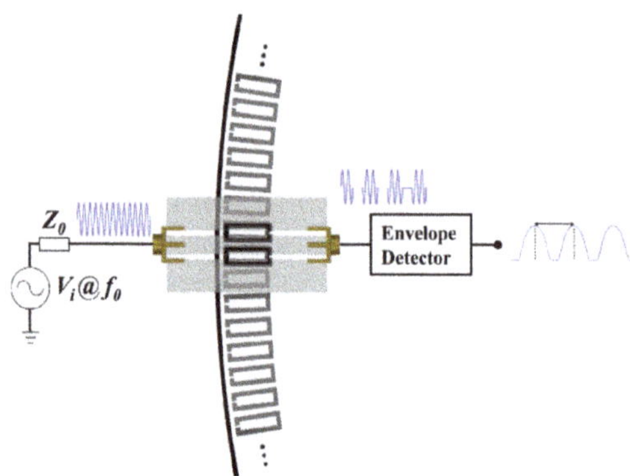

Figure 17. Sketch of the rotation sensor based on the edge configuration. From [29], reprinted with permission.

Figure 18 depicts the sketch of the sensor with a double chain of resonators to measure the angular speed, and an additional non-periodic chain to discriminate the motion direction. As it can be seen, two harmonic signals are needed in this case (injected through a combiner), one of them tuned to the resonance frequency of the resonators of the velocity chain, and the other one tuned to the resonance of the resonant elements of the direction chain. Rotor motion AM modulates each harmonic signal, and the corresponding envelope functions can be inferred by means of envelope detectors, after filtering each signal by means of a diplexer. Figure 19 shows the photograph of rotor (zoom view) and the stator, as well as a picture of the overall experimental set-up. The envelope functions inferred with such sensing system, corresponding to the angular velocity and direction detection, are shown in Figure 20 (note that the direction is detected by the increasing or decreasing sequence of pulses in the corresponding envelope function).

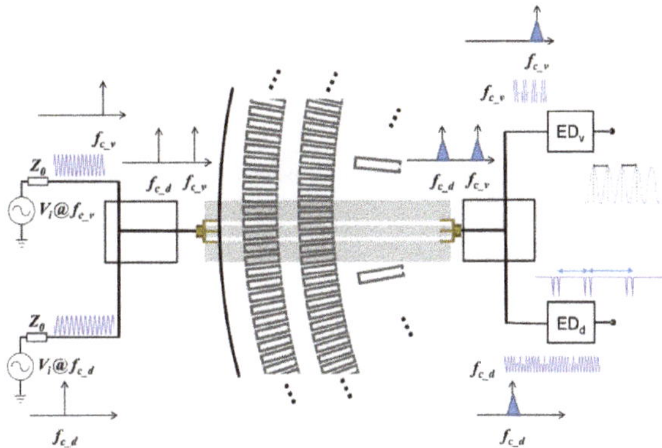

Figure 18. Sketch and working principle of the rotation sensor based on the edge configuration with two velocity chains and one direction detection chain. From [30], reprinted with permission.

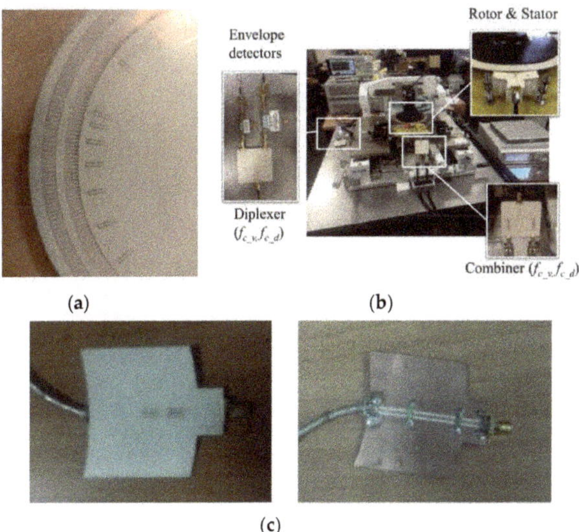

Figure 19. Photograph of the rotor (**a**), experimental set-up (**b**), and stator (**c**). From [30], reprinted with permission.

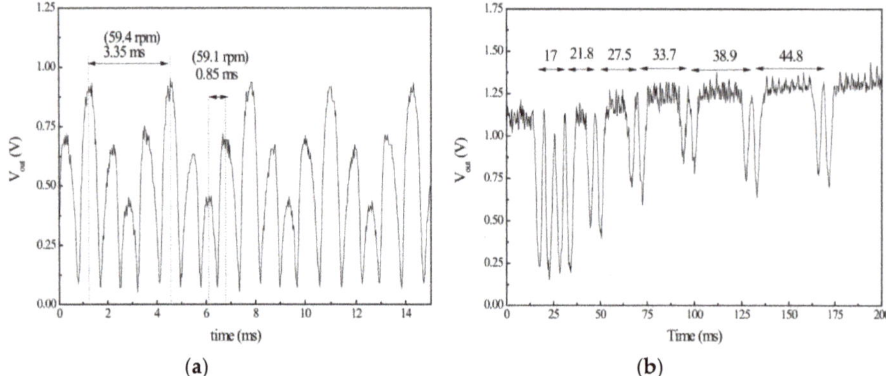

Figure 20. Envelope functions providing the angular velocity (**a**) and motion direction (**b**). The nominal rotary speed is 60 rpm. From [30], reprinted with permission.

In these angular displacement and velocity sensors based on the edge configuration, the number of cumulative pulses from a reference (REF) position provides the angle of rotation. Thus, the angular resolution (in degrees) is simply 360°/PPR, where PPR is the number of pulses per revolution. For the specific sensor system of Figure 19, with PPR = 1200, the angular resolution is thus 0.3°, a competitive value. Further details of this sensor (dimensions, considered substrates, specific components, and instrumentation) are given in [30]. A detailed analysis of angular velocity resolution and error for these sensors, out of the scope of this paper, can be found in [29]. As highlighted in [29,30], microwave rotary encoders represent a low-cost alternative to optical encoders. Moreover, microwave encoders can be of interest for operation in harsh and hostile environments (i.e., subjected to radiation, pollution, dirtiness, grease, etc.), where the optical counterparts may offer a limited robustness. Note that the dynamic range of these rotation sensors for angle measurement is theoretically unlimited, as these sensors are based on pulse counting. This is a clear advantage as compared to other sensors based on symmetry disruption and implemented by means of a single resonant element [21–23,26,27]. This unlimited dynamic range plus the achievable angular resolution situates these sensors as excellent candidates for angular displacement and velocity sensing.

4.3. Frequency Splitting Sensors

The operation principle of frequency splitting sensors is illustrated in Figure 21 [47]. A transmission line is symmetrically loaded with a pair of (not necessarily symmetric) resonators. If symmetry is preserved, a single notch in the transmission coefficient arises (provided the resonators are coupled to the line). Conversely, if symmetry is disrupted, two notches appear (splitting), and their relative distance depends on symmetry imbalance. Thus, this principle can be used for sensing. Although most frequency splitting sensors are implemented by symmetrically loading a line with a resonator pair (as indicated in Figure 21), in some implementations a cascade configuration, where the two resonant elements are placed at different locations in a transmission line, has been considered [56]. In this subsection, we report a frequency splitting sensor for dielectric characterization of liquids as an illustrative example, where a splitter/combiner configuration is used [58]. With such configuration, the effects of inter-resonator coupling (with negative impact on sensor sensitivity) can be avoided, as it was discussed in [57].

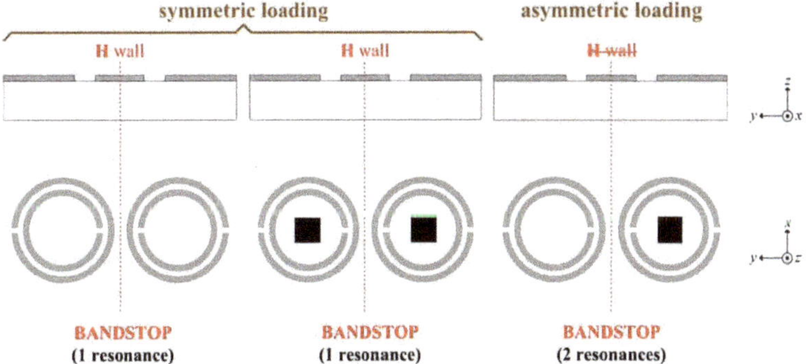

Figure 21. Illustration of frequency splitting through symmetry imbalance (considering square-shaped inclusions) in a CPW symmetrically loaded with a pair SRRs. From [47], reprinted with permission.

The topology of the designed sensor is shown in Figure 22a, whereas Figure 22b depicts a photograph (in perspective view) of the device, equipped with fluidic channels (device dimensions, substrate, as well as other details related to the fluidic channels and mechanical accessories can be found in [58]). Figure 23 depicts the frequency response for different configurations of channel loading, where the splitting effect for those cases of device imbalance can be observed. With the data of Figure 23 (resonance frequency splitting and difference in the notch magnitude, the output variables), a mathematical model following a scheme similar to the one reported in Section 4.1 can be inferred, provided the complex permittivity of mixtures of DI water, the reference sample, and ethanol can be obtained from independent sources. Such a model links the differential complex permittivity (real and imaginary parts), i.e., the input variables, to the differential output variables. Once the model is known, it can be used to determine the complex permittivity of liquid samples, by considering DI water as reference (REF) sample. In Figure 24, the measured responses of the device for different solutions of water–ethanol injected in the MUT channel, considering water as the REF liquid, are depicted. The inferred output variables are also represented in the figure. With these results and the mathematical model, the complex permittivity of such water–ethanol mixtures was obtained, and the results are depicted in Figure 25. It can be appreciated in that figure that the results lie within the limits of the Weiner model, and consequently these results validate the functionality of the sensor. Although these sensors belong to the category of symmetry-based sensors, and are therefore robust against the effects of ambient factors, the resolution and sensitivity of these sensors is, in general, not as good as the one of true differential sensors, as is discussed in the next section.

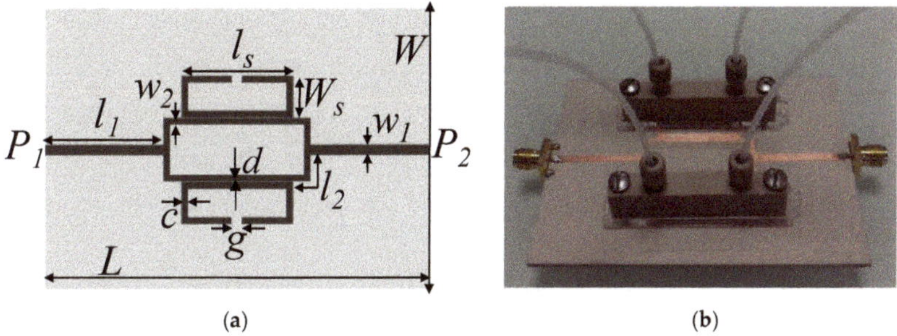

Figure 22. Layout (**a**) and perspective view (**b**) of the designed frequency-splitting sensor. From [58], reprinted with permission.

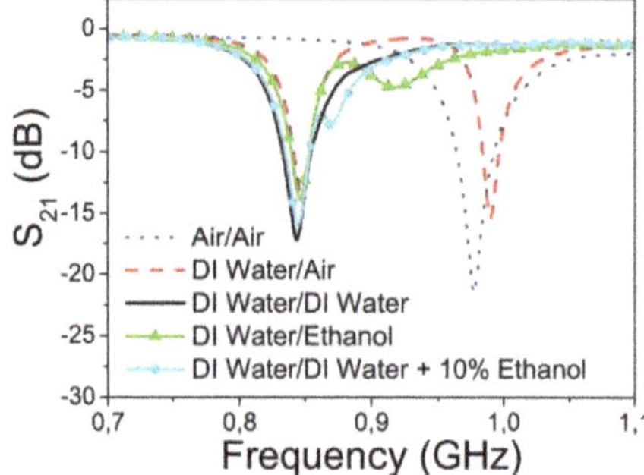

Figure 23. Measured responses of the sensor for different combinations of dielectric loads in the channels. From [58], reprinted with permission.

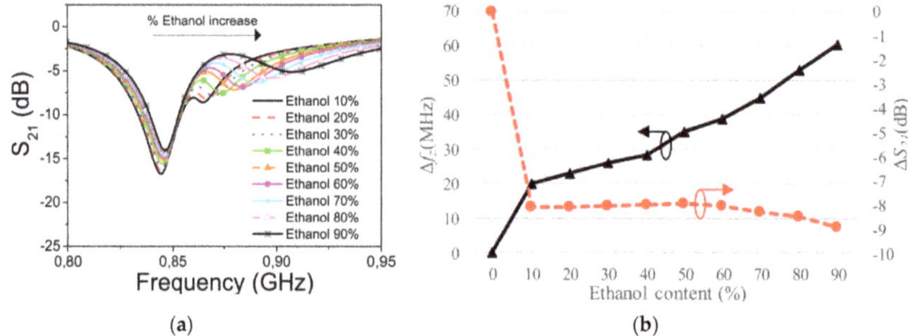

(a)   (b)

Figure 24. Measured frequency response of the sensor for various water–ethanol solutions (**a**) and representation of the resonance frequency splitting and notch magnitude difference, as a function of the ethanol concentration (**b**). From [58], reprinted with permission.

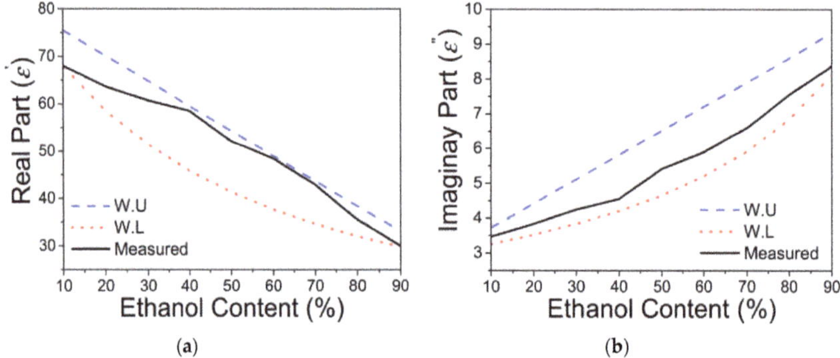

(a)   (b)

Figure 25. Real (**a**) and imaginary (**b**) part of the complex dielectric constant of water–ethanol solutions as a function of the ethanol content, inferred from the data of Figure 24 and the mathematical model developed in [58]. From [58], reprinted with permission.

4.4. Differential-Mode Sensors

Differential sensors based on a pair of uncoupled lines each one loaded with a resonant sensing element have been reported. The considered resonant elements are OCSRRs [62], SRRs [63], OSRRs [67,69], CSRRs [66], and DB-DGS resonators [64,65]. We report in this paper three prototype examples, including a sensor for complex permittivity measurement of liquids (based on DB-DGSs) [64], a sensor for the determination of volume fraction of solute in liquid solutions (based on OSRRs) [69], and a sensor devoted to the measurement of extremely small concentrations of electrolytes in DI water [66].

4.4.1. Measuring Electrolyte Concentration

Electrolytes are ions, such as sodium (Na^+), potassium (K^+), calcium (Ca^{2+}), chloride (Cl^-), and bicarbonate (HCO_3^-), among others, present in the blood and urine. Electrolytes play a key role in several vital functions, such as blood pH and pressure control, body hydration, nerve and muscle functions, etc. [95]. There are electrochemical methods to determine selectively the concentration of electrolytes in blood and urine, mainly based on the so-called ion-selective electrodes (ISE), but such methods are expensive. Measuring the total concentration of ions does not provide complete information, but it may be indicative of certain pathologies, or it can be useful as a prescreening diagnosis tool. Moreover, monitoring the real-time variations of the total electrolyte content in urine or blood samples, as compared to a reference (REF), is also of interest from a medical viewpoint. These functionalities can be achieved by means of the differential sensors reported in this section, where the main advantageous aspect is sensor cost (e.g., as compared to electrochemical sensors), and real-time functionality. Within this context, the development of low-cost microwave methods able to provide the electrolyte concentration in samples of blood and urine is of high interest. Nevertheless, to avoid using biosamples, and for the sake of easy comparisons, it is usual to estimate sensor performance by considering electrolyte solutions of standard liquid, typically DI water.

The reported differential sensor aimed to the measurement of electrolyte concentration in DI water is the one reported in [66], as it provides (to the best of our knowledge) the finest achieved resolution (0.125 g/L). Such a sensor is based on a pair of CSRR-loaded microstrip lines. The topology and a perspective view, including the fluidic channels, are depicted in Figure 26, whereas Figure 27 shows a photograph of the fabricated device. The output variable in these sensors is the cross-mode transmission coefficient, proportional to the difference between the transmission coefficient of both lines (the REF and the MUT line), namely,

$$S_{21}^{DC} = \frac{1}{2}(S_{21} - S_{43}) \tag{4}$$

This proportionality is valid provided the lines are uncoupled (otherwise, the cross-mode transmission coefficient is given by $S_{21}{}^{DC} = (1/2) \cdot (S_{21} + S_{23} - S_{41} - S_{43})$. The input variable is the electrolyte concentration of the MUT sample (provided pure DI water is the REF sample). The cross-mode transmission coefficient obtained after subsequently injecting solutions with different electrolyte content (with DI water in the REF channel) is depicted in Figure 28a. The considered electrolyte precursor is NaCl (which dissociates into anions and cations when dissolved in water). As the electrolyte concentration increases, the magnitude of the cross mode transmission coefficient shifts up. As it can be seen, a concentration as small as 0.125 g/L can be discriminated, as far as the corresponding curve is perfectly differentiated form the one of the REF sample (with 0 g/L of electrolyte concentration, i.e., pure DI water). Figure 28b depicts the dependence of the maximum cross-mode transmission coefficient with the electrolyte concentration. The maximum sensitivity (achieved for small electrolyte concentrations) is 0.034 $(g/L)^{-1}$, whereas the input dynamic range is at least 60 g/L. From the data points of Figure 28, the following calibration curve, with correlation coefficient of $R^2 = 0.9996$, results

$$NaCl\left(\frac{g}{L}\right) = 0.41e^{\left(\frac{\Delta S_{21}^{DC}}{4.2}\right)} + 4.91 \cdot 10^{-8} e^{\left(\frac{\Delta S_{21}^{DC}}{0.65}\right)} - 0.59 \tag{5}$$

This calibration curve can be used to determine unknown concentrations of NaCl in DI water, from the measured cross mode transmission coefficient. Note that other similar sensors for the measurement of electrolyte concentration have been reported [62–65]. However, the sensor of Figure 27 is the one exhibiting better resolution.

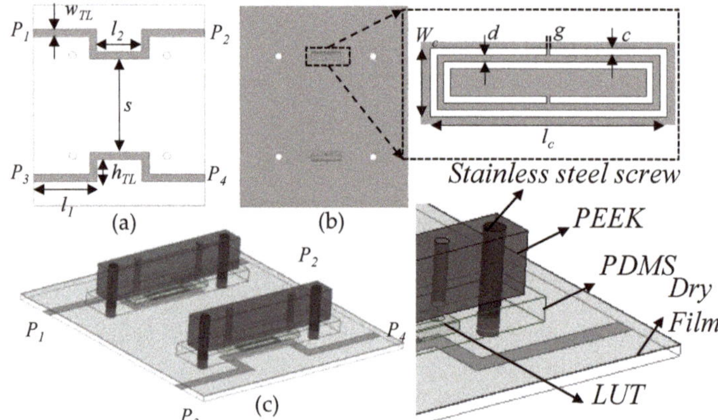

Figure 26. Topology (top view (**a**) and bottom view (**b**)) and 3D view (**c**) of the proposed CSRR-based sensor. From [66], reprinted with permission.

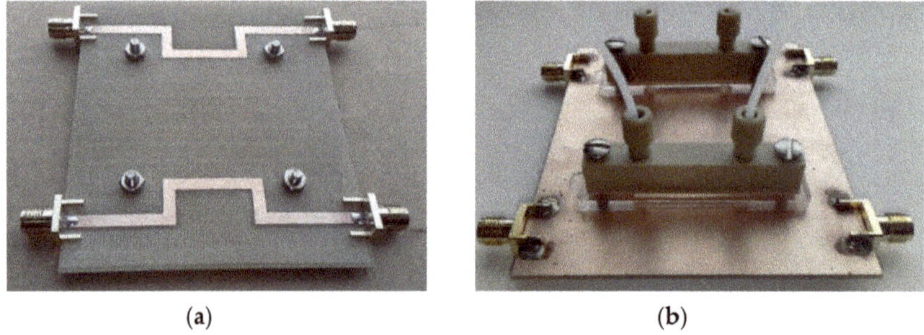

Figure 27. Perspective photographs of the CSRR-based differential sensor. (**a**) Top; (**b**) bottom. From [66], reprinted with permission.

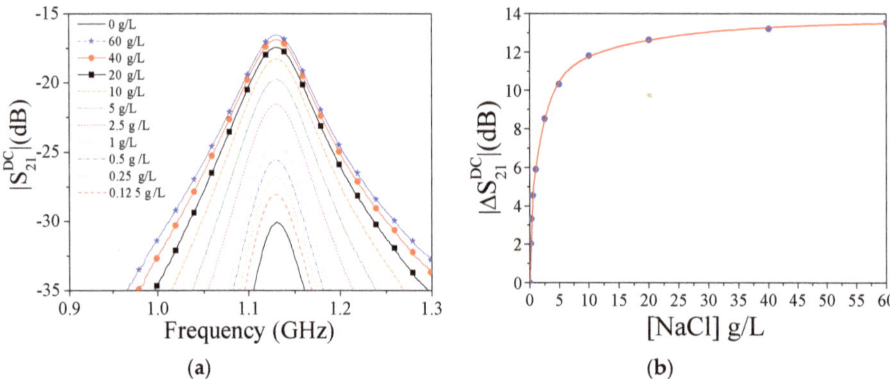

Figure 28. Measured cross-mode transmission coefficient for different NaCl concentrations (**a**) and dependence of the maximum incremental cross-mode transmission coefficient on the NaCl concentration (**b**). The calibration curve in (**b**) is included. From [66], reprinted with permission.

4.4.2. Complex Permittivity Measurement of Liquids

Determination of the complex permittivity of liquids from the measurement of the cross-mode transmission coefficient in differential-mode sensors was first reported in [62], where the sensing elements were OCSRRs. In this subsection, we report a similar sensor, but the sensing elements are DB-DGSs [64]. The device layout and photographs of the top and bottom view (including the fluidic channels) are depicted in Figure 29. The considered reference (REF) liquid is pure DI water, whereas different solutions of isopropanol in DI water constitute the MUT liquids. The measured cross mode transmission coefficients for the different solutions (where the volume fraction of isopropanol is the varying parameter) are depicted in Figure 30a. From this figure, the variation of the maximum of the cross mode transmission coefficient (magnitude) and the variation in the frequency where the maximum appears, as compared to the REF sample, are obtained. The results are depicted in Figure 30b. These are the two variables required to determine the complex permittivity of MUT liquids (consisting of the real and the imaginary parts). From the results of pure DI water, pure isopropanol, and a mixture of 50% isopropanol in DI water, a mathematical model linking the complex permittivity with the above-cited output variables was inferred. Then this model was used to obtain the complex dielectric constant of the different mixtures of isopropanol in DI water from the measured values of the output variable. The results are depicted in Figure 31, where it can be appreciated the good agreement with the results given by the Weiner model.

It is remarkable that the device is able to resolve volume fractions of at least 5%, i.e., better than the volume fraction resolved with the frequency splitting sensor reported in Section 4.3, and identical to the resolution of the differential sensor reported in [62]. It should be highlighted that the sensor of Figure 29 can be also used as single-ended sensor, and, in this case, the sensitivity is excellent, as discussed in [64]. Finally, let us emphasize that, as compared to the sensor reported in [62], in the sensor reported in this subsection, the fluidic channels are placed at the opposite side of the strip lines, thereby being simpler from the fabrication viewpoint.

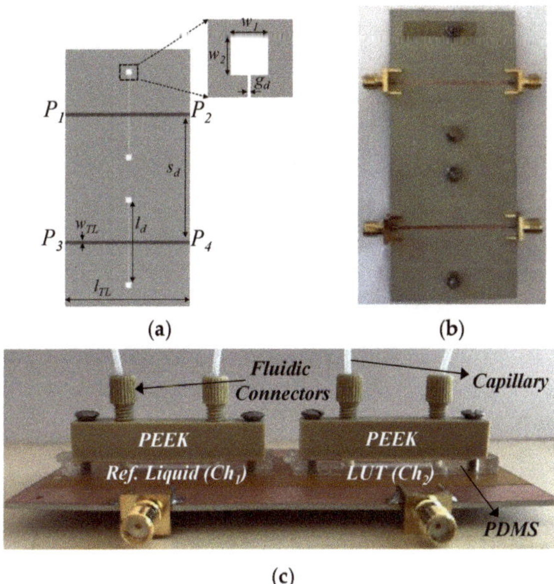

Figure 29. Layout (**a**) and photograph (top view (**b**) and 3D view (**c**)) of the DB-DGS-based sensor. From [64], reprinted with permission.

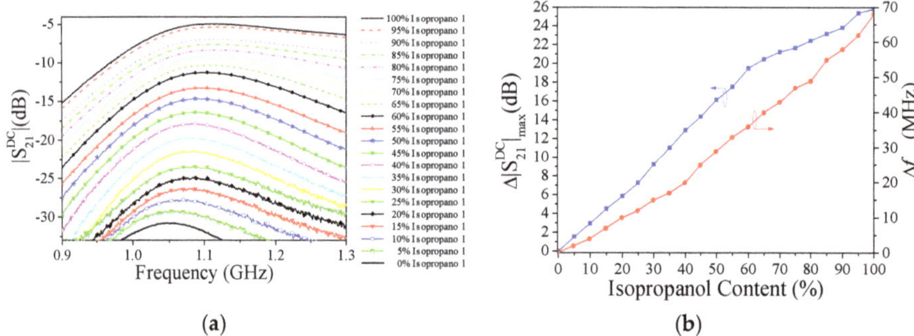

Figure 30. Measured cross-mode transmission coefficients for the different solutions of isopropanol in DI water (**a**) and representation of the output variables as a function of the isopropanol concentration (**b**). From [64], reprinted with permission.

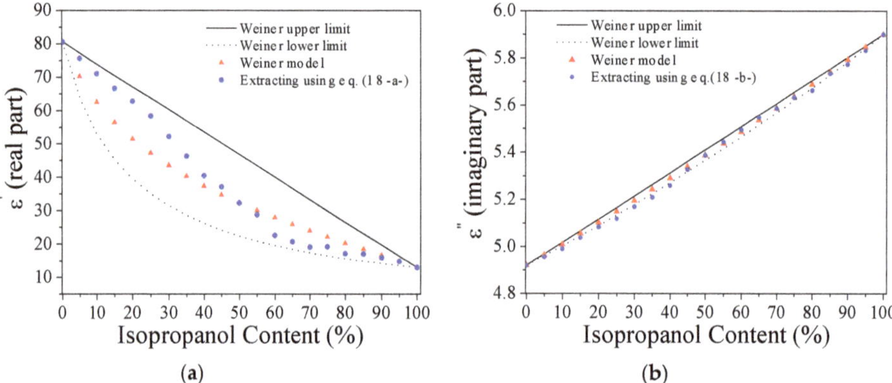

Figure 31. Real (**a**) and imaginary (**b**) parts of the complex dielectric constant for the different solutions of isopropanol in DI water. From [64], reprinted with permission.

4.4.3. Volume Fraction Determination in Liquid Solutions

Obviously, with the sensors reported in Sections 4.4.1 and 4.4.2, determination of the volume fraction in liquid solutions is possible. As this measurement involves only one input variable (the volume fraction), a single output variable suffices. In the sensor of Section 4.4.1, devoted to the measurement of electrolyte concentration in water, the output variable was the maximum of the cross mode transmission coefficient. The frequency position of such maximum slightly varies, but it is possible to obtain the value at a fixed frequency, particularly the resonance frequency of the resonant element. In this case, a single-frequency measurement suffices, and, in this case, the cross-mode transmission coefficient can be inferred by means of a reflective-mode differential measurement, involving a simple two-port measurement, as reported in [69]. The sensor device consists of two pairs of grounded OSRR-loaded lines connected to the coupled ports of a rat race coupler by means of two identical transmission line sections. The input signal is injected to the Δ-port, whereas the output port is the Σ-port (see the schematic in Figure 32).

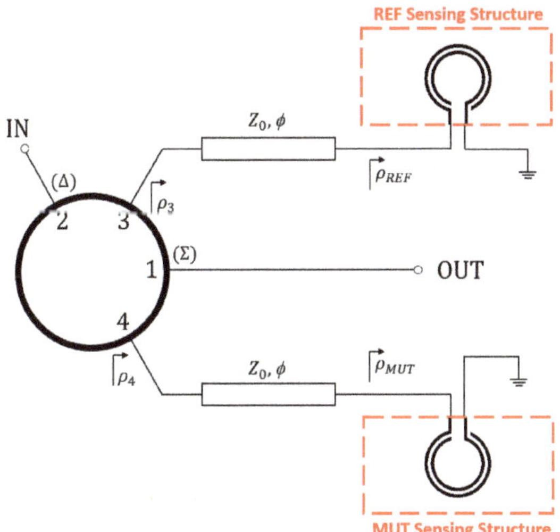

Figure 32. Schematic of the reflective-mode differential sensor based in OSRRs. From [69], reprinted with permission.

The transmission coefficient at f_0, the operating frequency of the rat-race coupler, in the structure of Figure 32 can be expressed as

$$S_{12} = -\frac{j}{2}e^{-j2\phi}\left(\rho_{REF} - \rho_{MUT}\right) \tag{6}$$

namely, S_{12} is proportional to the differential reflection coefficient. Therefore, by asymmetrically loading the sensing loads (OSRRs), it follows that differential sensing can be achieved. In Figure 33, the fabricated device, where microfluidic channels have been introduced in order to perform measurements of liquids, is shown. The objective of this sensor was the measurement of volume fraction of isopropanol in DI water. The measured transmission coefficients after subsequently injecting different solutions of isopropanol–water in the MUT channel, with DI water in the REF channel, are depicted in Figure 34. It can be seen that the transmission coefficient at f_0 experiences a strong variation as the concentration of isopropanol increases. Moreover, the sensor is able to resolve amounts of isopropanol content as small as 2.5%. Figure 35 depicts the dependence of the transmission coefficient at f_0 with the isopropanol concentration, including the calibration curve (with a correlation coefficient of $R^2 = 0.9987$), i.e.,

$$Fv(\%) = 184.942e^{\left(\frac{S_{21}(dB)}{5.687}\right)} + 81.618e^{\left(\frac{S_{21}(dB)}{45.449}\right)} - 34.793 \tag{7}$$

From this curve, the concentration of volume fraction of isopropanol can be inferred differentially, by means of a single two-port measurement.

The main differential aspect of the sensor of Figure 33 is the fact that a simple two-port measurement suffices to infer the differential output variable. Other differential sensor involving two-port measurements have been reported [12,68], but, in such sensors, neither resonant element is used, nor is operation in reflection mode considered. Nevertheless, note that in the recently presented sensor of [68], based on the phase difference generated in a pair of meandered lines, high sensitivity (11.52 dB/%) and resolution (1% of volume fraction) is obtained for the measurement of isopropanol content in DI water. This high sensitivity and resolution is in part due to the specific arrangement considered in [68], where a pair of rat race couplers is used (see further details in [68]). It is also important to highlight that in the sensor of Figure 33, a single rat-race coupler is used, because the sensor operates in reflection. By contrast in [68], the sensing device utilizes a pair of rat race couplers, with the consequently penalty in terms of size and cost.

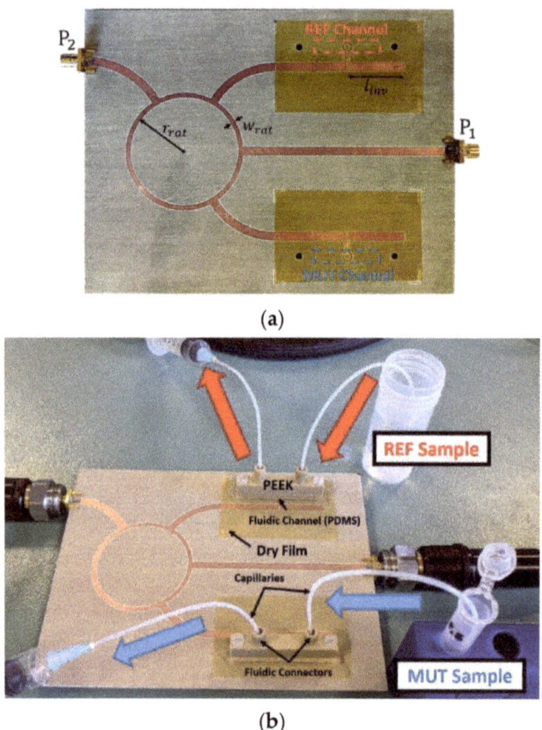

(a)

(b)

Figure 33. Top view of the sensor topology (**a**) and perspective view of the whole fabricated sensor (**b**). From [69], reprinted with permission.

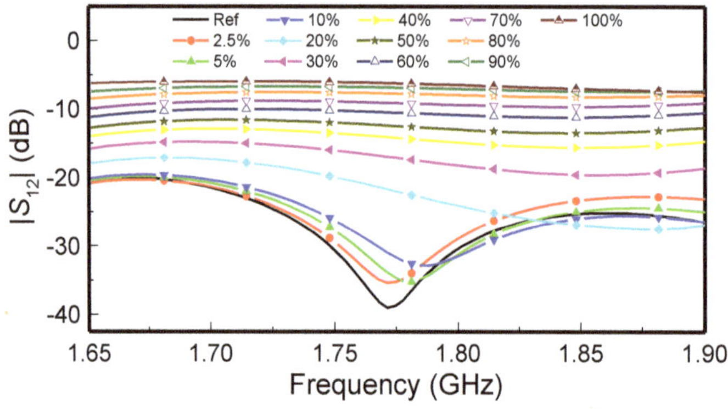

Figure 34. Frequency response of the sensing structure for different concentrations of isopropanol in DI water injected in the MUT channel, and pure DI water injected in the REF channel. From [69], reprinted with permission.

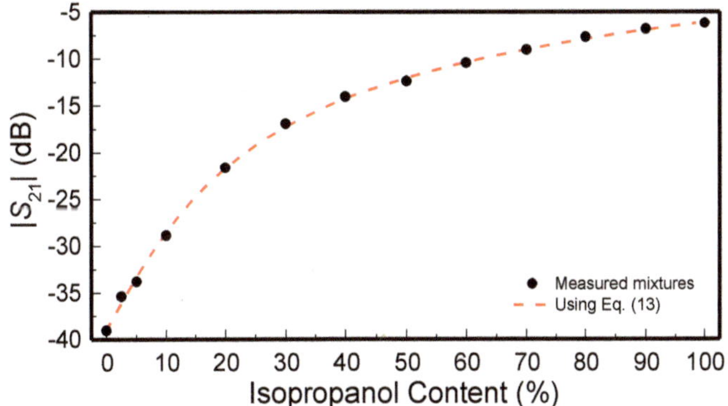

Figure 35. Variation of the transmission coefficient measured at f_0 (1.77GHz) with the isopropanol content, and calibration curve. From [69], reprinted with permission.

5. Conclusions

In conclusion, a review and recent developments of techniques for the implementation of planar microwave resonant sensors have been reported in this paper. These sensors have been classified in four groups, according to their working principle, i.e., frequency variation sensors, coupling modulation sensors, frequency splitting sensors, and differential-mode sensors. Their advantages and limitations have been pointed out and discussed. We have also presented the most usual electrically small resonant elements utilized for the implementation of sensors, and have provided some hints regarding the convenient resonator topologies for sensitivity and resolution optimization. Finally, we have illustrated the paper with some prototype sensors, mainly including sensors for dielectric characterization, but also sensors for measuring spatial variables.

Funding: This research activity was funded by MINECO Spain (project TEC2016-75650-R), by *Generalitat de Catalunya* (project 2017SGR-1159), and by FEDER (European Union) funds.

Acknowledgments: F.M. is in debt to *Institució Catalana de Recerca i Estudis Avançats* (ICREA) for supporting his work. J.M.E. acknowledges *Secreteraria d'Universitats i Recerca* (Gen. Cat.) and European Social Fund for the FI grant. P.V. acknowledges the *Juan de la Cierva Program* for supporting him through Project IJCI-2017-31339. M.G. acknowledges the *Universidad Politécnica de Madrid Young Researchers Support Program* (VJIDOCUPM18MGB) for its support.

Conflicts of Interest: The authors declare no conflicts of interest.

References

1. Gennarelli, G.; Romeo, S.; Scarfi, M.R.; Soldovieri, F. A microwave resonant sensor for concentration measurements of liquid solutions. *IEEE Sens. J.* **2013**, *13*, 1857–1864. [CrossRef]

2. Jha, A.K.; Akhtar, M.J. A generalized rectangular cavity approach for determination of complex permittivity of materials. *IEEE Trans. Instrum. Meas.* **2014**, *63*, 2632–2641. [CrossRef]

3. Wei, P.; Morey, B.; Dyson, T.; McMahon, N.; Hsu, Y.Y.; Gazman, S.; Klinker, L.; Ives, B.; Dowling, K.; Rafferty, C. A Conformal Sensor for Wireless Sweat Level Monitoring. In Proceedings of the IEEE Sensors, Baltimore, MD, USA, 3–6 November 2013.

4. Rodgers, M.M.; Pai, V.M.; Conroy, R.S. Recent advances in wearable sensors for health monitoring. *IEEE Sens. J.* **2014**, *15*, 3119–3126. [CrossRef]

5. Galindo-Romera, G.; Herraiz-Martínez, F.J.; Gil, M.; Martínez-Martínez, J.J.; Segovia-Vargas, D. Submersible printed split-ring resonator-based sensor for thin-film detection and permittivity characterization. *IEEE Sens. J.* **2016**, *16*, 3587–3596. [CrossRef]

6. Dehe, A.; Krozer, V.; Fricke, K.; Klingbeil, H.; Beilenhoff, K.; Hartnagel, H.L. Integrated microwave power sensor. *Electron. Lett.* **1995**, *31*, 2187–2188. [CrossRef]

7. Castillo-León, J.; Svendsen, W.E. *Lab-on-a-Chip Devices and Micro-Total Analysis Systems*; Springer: New York, NY, USA, 2015; ISBN 978-3-319-08686-6.

8. Grenier, K.; Dubuc, D.; Poleni, P.E.; Kumemura, M.; Toshiyoshi, H.; Fujii, T.; Fujita, H. Integrated broadband microwave and microfluidic sensor dedicated to bioengineering. *IEEE Trans. Microw. Theory Tech.* **2009**, *57*, 3246–3253. [CrossRef]

9. Chretiennot, T.; Dubuc, D.; Grenier, K. A microwave and microfluidic planar resonator for efficient and accurate complex permittivity characterization of aqueous solutions. *IEEE Trans. Microw. Theory Tech.* **2012**, *61*, 972–978. [CrossRef]

10. Salim, A.; Kim, S.H.; Park, J.Y.; Lim, S. Microfluidic Biosensor Based on Microwave Substrate-Integrated Waveguide Cavity Resonator. *J. Sens.* **2018**, *2018*, 1324145. [CrossRef]

11. Zarifi, M.H.; Sadabadi, H.; Hejazi, S.H.; Daneshmand, M.; Sanati-Nezhad, A. Noncontact and nonintrusive microwave-microfluidic flow sensor for energy and biomedical engineering. *Sci. Rep.* **2018**, *8*, 139. [CrossRef]

12. Damm, C.; Schüßler, M.; Puentes, M.; Maune, H.; Maasch, M.; Jakoby, R. Artificial transmission lines for high sensitive microwave sensors. In Proceedings of the IEEE Sensors, Christchurch, New Zealand, 25–28 October 2009; pp. 755–758.

13. Ferrández-Pastor, F.; García-Chamizo, J.; Nieto-Hidalgo, M. Electromagnetic differential measuring method: Application in microstrip sensors developing. *Sensors* **2017**, *17*, 1650. [CrossRef]

14. Muñoz-Enano, J.; Vélez, P.; Gil, M.; Martín, F. An Analytical Method to Implement High Sensitivity Transmission Line Differential Sensors for Dielectric Constant Measurements. *IEEE Sens. J.* **2019**, *20*, 178–184. [CrossRef]

15. Mandel, C.; Kubina, B.; Schüßler, M.; Jakoby, R. Passive chipless wireless sensor for two-dimensional displacement measurement. In Proceedings of the 41st European Microwave Conference, Manchester, UK, 9–14 October 2011; pp. 79–82.

16. Naqui, J.; Durán-Sindreu, M.; Martín, F. Novel sensors based on the symmetry properties of split ring resonators (SRRs). *Sensors* **2011**, *11*, 7545–7553. [CrossRef] [PubMed]

17. Naqui, J.; Durán-Sindreu, M.; Martín, F. On the symmetry properties of coplanar waveguides loaded with symmetric resonators: Analysis and potential applications. In Proceedings of the 2012 IEEE/MTT-S International Microwave Symposium Digest, Montreal, QC, Canada, 17–22 June 2012; pp. 1–3.

18. Naqui, J.; Durán-Sindreu, M.; Martín, F. Alignment and position sensors based on split ring resonators. *Sensors* **2012**, *12*, 11790–11797. [CrossRef]

19. Naqui, J.; Durán-Sindreu, M.; Martín, F. Transmission lines loaded with bisymmetric resonators and applications. In Proceedings of the IEEE MTT-S International Microwave Symposium Digest, Seattle, WA, USA, 7 June 2013; pp. 1–3.

20. Horestani, A.K.; Fumeaux, C.; Al-Sarawi, S.F.; Abbott, D. Displacement sensor based on diamond-shaped tapered split ring resonator. *IEEE Sens. J.* **2012**, *13*, 1153–1160. [CrossRef]

21. Horestani, A.K.; Abbott, D.; Fumeaux, C. Rotation sensor based on horn-shaped split ring resonator. *IEEE Sens. J.* **2013**, *13*, 3014–3015. [CrossRef]

22. Naqui, J.; Martı, F. Transmission lines loaded with bisymmetric resonators and their application to angular displacement and velocity sensors. *IEEE Trans. Microw. Theory Tech.* **2013**, *61*, 4700–4713. [CrossRef]

23. Ebrahimi, A.; Withayachumnankul, W.; Al-Sarawi, S.F.; Abbott, D. Metamaterial-inspired rotation sensor with wide dynamic range. *IEEE Sens. J.* **2014**, *14*, 2609–2614. [CrossRef]

24. Horestani, A.K.; Naqui, J.; Shaterian, Z.; Abbott, D.; Fumeaux, C.; Martín, F. Two-dimensional alignment and displacement sensor based on movable broadside-coupled split ring resonators. *Sens. Actuators A* **2014**, *210*, 18–24. [CrossRef]

25. Horestani, A.K.; Naqui, J.; Abbott, D.; Fumeaux, C.; Martín, F. Two-dimensional displacement and alignment sensor based on reflection coefficients of open microstrip lines loaded with split ring resonators. *Electron. Lett.* **2014**, *50*, 620–622. [CrossRef]

26. Naqui, J.; Martín, F. Angular displacement and velocity sensors based on electric-LC (ELC) loaded microstrip lines. *IEEE Sens. J.* **2013**, *14*, 939–940. [CrossRef]

27. Naqui, J.; Coromina, J.; Karami-Horestani, A.; Fumeaux, C.; Martín, F. Angular displacement and velocity sensors based on coplanar waveguides (CPWs) loaded with S-shaped split ring resonators (S-SRR). *Sensors* **2015**, *15*, 9628–9650. [CrossRef] [PubMed]

28. Naqui, J.; Martín, F. Application of broadside-coupled split ring resonator (BC-SRR) loaded transmission lines to the design of rotary encoders for space applications. In Proceedings of the IEEE MTT-S International Microwave Symposium, San Francisco, CA, USA, 22–27 May 2016; pp. 1–4.

29. Mata-Contreras, J.; Herrojo, C.; Martín, F. Application of split ring resonator (SRR) loaded transmission lines to the design of angular displacement and velocity sensors for space applications. *IEEE Trans. Microw. Theory Tech.* **2017**, *65*, 4450–4460. [CrossRef]

30. Mata-Contreras, J.; Herrojo, C.; Martín, F. Detecting the rotation direction in contactless angular velocity sensors implemented with rotors loaded with multiple chains of resonators. *IEEE Sens. J.* **2018**, *18*, 7055–7065. [CrossRef]

31. Puentes, M. *Planar Metamaterial Based Microwave Sensor Arrays for Biomedical Analysis and Treatment*; Springer: Heidelberg, Germany, 2014; ISBN 978-3319060408.

32. Ebrahimi, A.; Withayachumnankul, W.; Al-Sarawi, S.; Abbott, D. High-sensitivity metamaterial-inspired sensor for microfluidic dielectric characterization. *IEEE Sens. J.* **2013**, *14*, 1345–1351. [CrossRef]

33. Schueler, M.; Mandel, C.; Puentes, M.; Jakoby, R. Metamaterial inspired microwave sensors. *IEEE Microw. Mag.* **2012**, *13*, 57–68. [CrossRef]

34. Boybay, M.S.; Ramahi, O.M. Material characterization using complementary split-ring resonators. *IEEE Trans. Instrum. Meas.* **2012**, *61*, 3039–3046. [CrossRef]

35. Lee, C.S.; Yang, C.L. Complementary split-ring resonators for measuring dielectric constants and loss tangents. *IEEE Microw. Wirel. Compon. Lett.* **2014**, *24*, 563–565. [CrossRef]

36. Yang, C.L.; Lee, C.S.; Chen, K.W.; Chen, K.Z. Noncontact measurement of complex permittivity and thickness by using planar resonators. *IEEE Trans. Microw. Theory Tech.* **2015**, *64*, 247–257. [CrossRef]

37. Withayachumnankul, W.; Jaruwongrungsee, K.; Tuantranont, A.; Fumeaux, C.; Abbott, D. Metamaterial-based microfluidic sensor for dielectric characterization. *Sens. Actuators A* **2013**, *189*, 233–237. [CrossRef]

38. Salim, A.; Lim, S. Complementary split-ring resonator-loaded microfluidic ethanol chemical sensor. *Sensors* **2016**, *16*, 1802. [CrossRef]

39. Su, L.; Mata-Contreras, J.; Vélez, P.; Fernández-Prieto, A.; Martín, F. Analytical method to estimate the complex permittivity of oil samples. *Sensors* **2018**, *18*, 984. [CrossRef] [PubMed]

40. Abdolrazzaghi, M.; Zarifi, M.H.; Daneshmand, M. Sensitivity enhancement of split ring resonator based liquid sensors. In Proceedings of the IEEE Sensors, 30 October–3 November 2016.

41. Abdolrazzaghi, M.; Zarifi, M.H.; Pedrycz, W.; Daneshmand, M. Robust ultra-high resolution microwave planar sensor using fuzzy neural network approach. *IEEE Sens. J.* **2016**, *17*, 323–332. [CrossRef]

42. Zarifi, M.H.; Daneshmand, M. Monitoring solid particle deposition in lossy medium using planar resonator sensor. *IEEE Sens. J.* **2017**, *17*, 7981–7989. [CrossRef]

43. Zarifi, M.H.; Deif, S.; Abdolrazzaghi, M.; Chen, B.; Ramsawak, D.; Amyotte, M.; Vahabisani, N.; Hashisho, Z.; Chen, W.; Daneshmand, M. A microwave ring resonator sensor for early detection of breaches in pipeline coatings. *IEEE Trans. Ind. Electron.* **2017**, *65*, 1626–1635. [CrossRef]

44. Abdolrazzaghi, M.; Daneshmand, M.; Iyer, A.K. Strongly enhanced sensitivity in planar microwave sensors based on metamaterial coupling. *IEEE Trans. Microw. Theory Tech.* **2018**, *66*, 1843–1855. [CrossRef]

45. Su, L.; Mata-Contreras, J.; Vélez, P.; Martín, F. Estimation of the Complex Permittivity of Liquids by means of Complementary Split Ring Resonator (CSRR) Loaded Transmission Lines. In Proceedings of the IEEE MTT-S International Microwave Workshop Series on Advanced Materials and Processes, Pavia, Italy, 20–22 September 2017; pp. 1–3.

46. Ebrahimi, A.; Scott, J.; Ghorbani, K. Ultrahigh-Sensitivity Microwave Sensor for Microfluidic Complex Permittivity Measurement. *IEEE Trans. Microw. Theory Tech.* **2019**, *67*, 4269–4277. [CrossRef]

47. Naqui, J. *Symmetry Properties in Transmission Lines Loaded with Electrically Small Resonators: Circuit Modeling and Applications*; Springer: Heidelberg, Germany, 2016; ISBN 978-3-319-24566-9.

48. Naqui, J.; Martín, F. Microwave sensors based on symmetry properties of resonator-loaded transmission lines. *J. Sens.* **2015**, *2015*, 741853. [CrossRef]

49. Herrojo, C.; Muela, F.J.; Mata-Contreras, J.; Paredes, F.; Martín, F. High-Density Microwave Encoders for Motion Control and Near-Field Chipless-RFID. *IEEE Sens. J.* **2019**, *19*, 3673–3682. [CrossRef]

50. Herrojo, C.; Paredes, F.; Martín, F. Double-stub loaded microstrip line reader for very high data density microwave encoders. *IEEE Trans. Microw. Theory Tech.* **2019**, *67*, 3527–3536. [CrossRef]

51. Naqui, J.; Damm, C.; Wiens, A.; Jakoby, R.; Su, L.; Martín, F. Transmission lines loaded with pairs of magnetically coupled stepped impedance resonators (SIRs): Modeling and application to microwave sensors. In Proceedings of the IEEE MTT-S International Microwave Symposium, Tampa, FL, USA, 1–6 June 2014; pp. 1–4.

52. Su, L.; Naqui, J.; Mata-Contreras, J.; Martín, F. Modeling metamaterial transmission lines loaded with pairs of coupled split-ring resonators. *IEEE Antennas Wirel. Propag. Lett.* **2014**, *14*, 68–71. [CrossRef]

53. Su, L.; Naqui, J.; Mata, J.; Martín, F. Dual-band epsilon-negative (ENG) transmission line metamaterials based on microstrip lines loaded with pairs of coupled complementary split ring resonators (CSRRs): Modeling, analysis and applications. In Proceedings of the 9th International Congress on Advanced Electromagnetic Materials in Microwaves and Optics, Metamaterials 2015, Oxford, UK, 7–12 September 2015; pp. 298–300.

54. Su, L.; Naqui, J.; Mata-Contreras, J.; Vélez, P.; Martín, F. Transmission line metamaterials based on pairs of coupled split ring resonators (SRRs) and complementary split ring resonators (CSRR): A comparison to the light of the lumped element equivalent circuits. In Proceedings of the International Conference on Electromagnetics for Advanced Applications, ICEAA 2015, Torino, Italy, 7–11 September 2015; pp. 891–894.

55. Su, L.; Naqui, J.; Mata-Contreras, J.; Martín, F. Modeling and applications of metamaterial transmission lines loaded with pairs of coupled complementary split-ring resonators (CSRRs). *IEEE Antennas Wirel. Propag. Lett.* **2015**, *15*, 154–157. [CrossRef]

56. Naqui, J.; Damm, C.; Wiens, A.; Jakoby, R.; Su, L.; Mata-Contreras, J.; Martín, F. Transmission lines loaded with pairs of stepped impedance resonators: Modeling and application to differential permittivity measurements. *IEEE Trans. Microw. Theory Tech.* **2016**, *64*, 3864–3877. [CrossRef]

57. Su, L.; Mata-Contreras, J.; Vélez, P.; Martín, F. Splitter/combiner microstrip sections loaded with pairs of complementary split ring resonators (CSRRs): Modeling and optimization for differential sensing applications. *IEEE Trans. Microw. Theory Tech.* **2016**, *64*, 4362–4370. [CrossRef]

58. Vélez, P.; Su, L.; Grenier, K.; Mata-Contreras, J.; Dubuc, D.; Martín, F. Microwave microfluidic sensor based on a microstrip splitter/combiner configuration and split ring resonators (SRRs) for dielectric characterization of liquids. *IEEE Sens. J.* **2017**, *17*, 6589–6598. [CrossRef]

59. Ebrahimi, A.; Scott, J.; Ghorbani, K. Differential sensors using microstrip lines loaded with two split-ring resonators. *IEEE Sens. J.* **2018**, *18*, 5786–5793. [CrossRef]

60. Gil, M.; Vélez, P.; Aznar-Ballesta, F.; Muñoz-Enano, J.; Martín, F. Differential Sensor based on Electro-Inductive Wave (EIW) Transmission Lines for Dielectric Constant Measurements and Defect Detection. *IEEE Trans. Antennas Propag.* **2020**, *68*, 1876–1886. [CrossRef]

61. Shi, D.; Guo, J.; Chen, L.; Xia, C.; Yu, Z.; Ai, Y.; Li, C.M.; Kang, Y.; Wang, Z. Differential microfluidic sensor on printed circuit board for biological cells analysis. *Electrophoresis* **2015**, *36*, 1854–1858. [CrossRef]

62. Velez, P.; Grenier, K.; Mata-Contreras, J.; Dubuc, D.; Martín, F. Highly-sensitive microwave sensors based on open complementary split ring resonators (OCSRRs) for dielectric characterization and solute concentration measurement in liquids. *IEEE Access* **2018**, *6*, 48324–48338. [CrossRef]

63. Vélez, P.; Muñoz-Enano, J.; Grenier, K.; Mata-Contreras, J.; Dubuc, D.; Martín, F. Split Ring Resonator-Based Microwave Fluidic Sensors for Electrolyte Concentration Measurements. *IEEE Sens. J.* **2018**, *19*, 2562–2569. [CrossRef]

64. Vélez, P.; Muñoz-Enano, J.; Gil, M.; Mata-Contreras, J.; Martín, F. Differential Microfluidic Sensors Based on Dumbbell-Shaped Defect Ground Structures in Microstrip Technology: Analysis, Optimization, and Applications. *Sensors* **2019**, *19*, 3189. [CrossRef]

65. Muñoz-Enano, J.; Vélez, P.; Gil, M.; Mata-Contreras, J.; Martín, F. Microwave comparator based on defect ground structures. In Proceedings of the European Microwave Conference in Central Europe, Prague, Czech Republic, 13–15 May 2019.

66. Vélez, P.; Muñoz-Enano, J.; Martín, F. Electrolyte concentration measurements in DI water with 0.125 g/L resolution by means of CSRR-based structures. In Proceedings of the 49th European Microwave Conference, Paris, France, 1–3 October 2019.

67. Muñoz-Enano, J.; Vélez, P.; Gil, M.; Mata-Contreras, J.; Grenier, K.; Dubuc, D.; Martín, F. Microstrip Lines Loaded with Metamaterial-Inspired Resonators for Microwave Sensors/Comparators with Optimized Sensitivity. In Proceedings of the 49th European Microwave Conference, Paris, France, 1–3 October 2019; pp. 754–757.

68. Muñoz-Enano, J.; Vélez, P.; Gil, M.; Martín, F. Differential-mode to common-mode conversion detector based on rat-race couplers: Analysis and application to microwave sensors and comparators. *IEEE Trans. Microw. Theory Tech.* **2020**, *68*, 1312–1325. [CrossRef]

69. Muñoz-Enano, J.; Vélez, P.; Gil, M.; Martín, F. Microfluidic reflective-mode differential sensor based on open split ring resonators (OSRRs). *Int. J. Microw. Wirel. Technol.* **2020**. to be published.

70. Ebrahimi, A.; Scott, J.; Ghorbani, K. Transmission Lines Terminated with LC Resonators for Differential Permittivity Sensing. *IEEE Microw. Wirel. Compon. Lett.* **2018**, *28*, 1149–1151. [CrossRef]

71. Ebrahimi, A.; Scott, J.; Ghorbani, K. Microwave reflective biosensor for glucose level detection in aqueous solutions. *Sens. Actuators A* **2019**, *301*, 111662. [CrossRef]

72. Vélez, P.; Muñoz-Enano, J.; Martín, F. Differential Sensing Based on Quasi-Microstrip Mode to Slot-Mode Conversion. *IEEE Microw. Wirel. Compon. Lett.* **2019**, *29*, 690–692. [CrossRef]

73. Marqués, R.; Martin, F.; Sorolla, M. *Metamaterials with Negative Parameters: Theory, Design, and Microwave Applications*; John Wiley & Sons: Hoboken, NJ, USA, 2013; ISBN 978-0-471-74582-2.

74. Martín, F. *Artificial Transmission Lines for RF and Microwave Applications*; John Wiley & Sons: Hoboken, NJ, USA, 2015; ISBN 9781119058403.

75. Durán-Sindreu, M.; Naqui, J.; Paredes, F.; Bonache, J.; Martín, F. Electrically small resonators for planar metamaterial, microwave circuit and antenna design: A comparative analysis. *Appl. Sci.* **2012**, *2*, 375–395. [CrossRef]

76. Makimoto, M.; Yamashita, S. Compact bandpass filters using stepped impedance resonators. *Proc. IEEE* **1979**, *67*, 16–19. [CrossRef]

77. Naqui, J.; Durán-Sindreu, M.; Bonache, J.; Martín, F. Implementation of shunt-connected series resonators through stepped-impedance shunt stubs: Analysis and limitations. *IET Microw. Antennas Propag.* **2011**, *5*, 1336–1342. [CrossRef]

78. Pendry, J.B.; Holden, A.J.; Robbins, D.J.; Stewart, W.J. Magnetism from conductors and enhanced nonlinear phenomena. *IEEE Trans. Microw. Theory Tech.* **1999**, *47*, 2075–2084. [CrossRef]

79. Martel, J.; Marqués, R.; Falcone, F.; Baena, J.D.; Medina, F.; Martín, F.; Sorolla, M. A new LC series element for compact bandpass filter design. *IEEE Microw. Wirel. Compon. Lett.* **2004**, *14*, 210–212. [CrossRef]

80. Schurig, D.; Mock, J.J.; Smith, D.R. Electric-field-coupled resonators for negative permittivity metamaterials. *Appl. Phys. Lett.* **2006**, *88*, 041109. [CrossRef]

81. Falcone, F.; Lopetegi, T.; Baena, J.D.; Marqués, R.; Martín, F.; Sorolla, M. Effective negative-ε stop-band microstrip lines based on complementary split ring resonators. *IEEE Microw. Wirel. Compon. Lett.* **2004**, *14*, 280–282. [CrossRef]

82. Velez, A.; Aznar, F.; Bonache, J.; Velazquez-Ahumada, M.C.; Martel, J.; Martin, F. Open complementary split ring resonators (OCSRRs) and their application to wideband CPW band pass filters. *IEEE Microw. Wirel. Compon. Lett.* **2009**, *19*, 197–199. [CrossRef]

83. Ahn, D.; Park, J.S.; Kim, C.S.; Kim, J.; Qian, Y.; Itoh, T. A design of the low-pass filter using the novel microstrip defected ground structure. *IEEE Trans. Microw. Theory Tech.* **2001**, *49*, 86–93. [CrossRef]

84. Naqui, J.; Durán-Sindreu, M.; Martín, F. Differential and single-ended microstrip lines loaded with slotted magnetic-LC resonators. *Int. J. Antennas Propag.* **2013**, *2013*. [CrossRef]

85. Chen, H.; Ran, L.; Huangfu, J.; Zhang, X.; Chen, K.; Grzegorczyk, T.M.; Au Kong, J. Left-handed materials composed of only S-shaped resonators. *Physical Review E* **2004**, *70*, 057605. [CrossRef]

86. Baena, J.D.; Marqués, R.; Medina, F.; Martel, J. Artificial magnetic metamaterial design by using spiral resonators. *Phys. Rev. B* **2004**, *69*, 014402. [CrossRef]

87. Marqués, R.; Baena, J.D.; Martel, J.; Medina, F.; Falcone, F.; Sorolla, M.; Martín, F. Novel small resonant electromagnetic particles for metamaterial and filter design. In Proceedings of the Electromagnetics in Advanced Applications, Torino, Italy, 8–12 September 2003; pp. 439–442.

88. Marqués, R.; Medina, F.; Rafii-El-Idrissi, R. Role of bianisotropy in negative permeability and left-handed metamaterials. *Phys. Rev. B* **2002**, *65*, 144440. [CrossRef]

89. Vélez, A.; Aznar, F.; Durán-Sindreu, M.; Bonache, J.; Martin, F. Open complementary split ring resonators (OCSRRs): The missing particle. In Proceedings of the Metamaterials 2009, London, UK, 30 August–4 September 2009.

90. Muñoz-Enano, J.; Vélez, P.; Herrojo, C.; Gil, M.; Martín, F. On the Sensitivity of Microwave Sensors based on Slot Resonators and Frequency Variation. In Proceedings of the International Conference on Electromagnetics in Advanced Applications (ICEAA), Granada, Spain, 9–13 September 2019; pp. 0112–0115.
91. Bao, J.Z.; Swicord, M.L.; Davis, C.C. Microwave dielectric characterization of binary mixtures of water, methanol, and ethanol. *J. Chem. Phys.* **1996**, *104*, 4441–4450. [CrossRef]
92. Eitel, E. Basics of rotary encoders: Overview and new technologies. *Mach. Des. Mag.* **2014**, 7.
93. McMillan, G.K.; Considine, D.M. *Process Instruments and Controls Handbook*, 5th ed.; McGraw Hill: New York, NY, USA, 1999; ISBN 978-0-07-012582-7.
94. Li, X.; Qi, J.; Zhang, Q.; Zhang, Y. Bias-tunable dual-mode ultraviolet photodetectors for photoelectric tachometer. *Appl. Phys. Lett.* **2014**, *104*, 041108. [CrossRef]
95. Hogan, M.A. *Pearson Reviews & Rationales: Fluids, Electrolytes, and Acid-Base Balance with Nursing Reviews & Rationales*, 4th ed.; Pearson: London, UK, 2017; ISBN 978-0134457710.

Article

Multilayered Balanced Dual-Band Bandpass Filter Based on Magnetically Coupled Open-Loop Resonators with Intrinsic Common-Mode Rejection

Jose L. Medran del Rio [1], Aintzane Lujambio [2], Armando Fernández-Prieto [1,*], Alejandro Javier Martinez-Ros [3], Jesús Martel [4] and Francisco Medina [1]

[1]   Departamento de Electrónica y Electromagnetismo, Facultad de Física, Universidad de Sevilla, 41012 Sevilla, Spain; jomedel@us.es (J.L.M.d.R.); armandof@us.es (A.F.-P.); medina@us.es (F.M.)
[2]   Parts Laboratory Department, Alter Technology TÜV Nord S.A.U., 41092 Seville, Spain; aintzane.lujambio@altertechnology.com
[3]   Departamento de Física Aplicada I, Universidad de Sevilla ETSII, 41012 Sevilla, Spain; amartinez49@us.es
[4]   Departamento de Física Aplicada II, Universidad de Sevilla ETSA, 41012 Sevilla, Spain; martel@us.es
*   Correspondence: armandof@us.es

Received: 27 March 2020; Accepted: 27 April 2020; Published: 29 April 2020

Abstract: A new dual-band balanced bandpass filter based on magnetically coupled open-loop resonators in multilayer technology is proposed in this paper. The lower differential passband, centered at the Global Positioning System (GPS) L1 frequency, 1.575 GHz, was created by means of two coupled resonators etched in the middle layer of the structure, while the upper differential passband, centered at a Wi-Fi frequency of 2.4 GHz, was generated by coupling two resonators on the top layer. Magnetic coupling was used to design both passbands, leading to an intrinsic common-mode rejection of 39 dB within the lower passband and 33 dB within the upper passband. Simulation and measurement results are provided to verify the usefulness of the proposed dual-band differential bandpass filter.

Keywords: dual-band differential filter; common-mode suppression; magnetic coupling; multilayer structure

---

1. Introduction

In recent years, the use of differential signals has gained increasing attention for both digital high-speed and analog microwave circuit applications [1,2]. This interest in differential devices is mainly due to their higher immunity to environmental noise, better electromagnetic compatibility, lower level of electromagnetic interference (EMI) and better signal to noise ratio performance, when compared to their single-ended counterparts. Despite all these advantages, differential devices can suffer from the presence of common-mode (CM) noise, mainly caused by amplitude unbalance and time skew of the differential signals. Therefore, to ensure the integrity of the differential signal over the frequency range of interest, strong CM rejection is highly desired. Different types of microwave devices in their differential version can be found nowadays in the literature: power dividers and combiners [3–6], diplexers [7] or passive equalizers [8]. However, differential-mode balanced bandpass filters (DM-BPFs) are, undoubtedly, the devices that have attracted the most attention in the literature. DM-BPFs with single/multiple differential passbands are required to have good differential-mode (DM) transmission within the passbands (low insertion loss, IL), good out-of-band rejection (high selectivity) and high CM rejection level (at least within the differential passbands). During the last decade, much effort has been focused on the design of balanced single-band BPFs, in such a away that a huge number of works can be found in the specialized literature. To mention just a few, see for example, [2,9–30] and the references therein. Nevertheless, much less research has been done on balanced dual-band BPFs [27–42] when compared with single-ended versions. It is well known that balanced dual-band BPFs are key components for multi-band systems operating under several wireless standards from Institute of Electrical and Electronics Engineers (IEEE), such as IEEE 802.11 (Wi-Fi 2.45/5.3 GHz) and IEEE 802.16 (WiMAX - 2.45/3.5 GHz), as well as for Global Navigation Satellite System (GNSS) systems such as GPS (L1/L2 - 1.575/1.227 GHz) [43]. Regarding the methods developed to design balanced dual-band BPFs, the use of electrically coupled resonators is by far the most commonly used strategy, usually leading to good common-mode rejection. For example, in [27] a fourth-order dual-band balanced BPF based on half-wavelength resonators is presented. Although the filter offers good DM performance and good CM noise suppression, it is difficult to control the bandwidths of the two differential-passbands independently. In addition, it requires the introduction of lumped-elements to properly reject the common-mode signal since for the unloaded resonators CM rejection is poor. In [28,31–34], stepped impedance resonators (SIRs) are used to perform balanced dual-band operation. The impedance and length ratios of the SIRs were adjusted to realize the desired dual-band response. However, since the level of CM noise rejection for electrically coupled SIRs is relatively poor, additional elements such as lumped inductors/capacitors/resistors [28], common-mode rejection stages based on differential lines with defected ground structures (DGS) [31,32], or open-circuit stubs [33,34] must be added to the resonators in order to improve their CM performance. In all the aforementioned cases, both DM and CM exhibit good performance, but at the expense of increasing the filter size and the complexity of the design process, since two different devices must be designed: the filter itself and the additional elements to suppress common-mode transmission.

As in the cases exposed in [31,32], DGSs have also been used in [29,35] to improve CM rejection. In those papers, a multilayer structure was proposed with the ground plane located in the middle. By introducing a slotline in the ground plane, common-mode transmission is avoided since the location of the slot is not compatible with the pattern of common-mode currents. In all these cases [29,31,32,35], the use of slotted ground plane provides good CM rejection at the expense of increasing radiation losses and making difficult system integration (for many practical applications a solid ground plane is required).

Lumped or quasi-lumped microwave resonators are not the only structures used to design balanced dual-band bandpass filters. Distributed structures have also been used for this purpose. Some works based on them can be found in the literature [30,36–40]. For example, asymmetrical coupled lines [36] and substrate integrated waveguide (SIW) technology [37] are interesting recent approaches for balanced dual-band BPF design. These designs provide both good DM and CM response, but typically suffer from the problem of large electrical size. Moreover, both [36,37] require the use of a large number of via-holes, which makes their performance very dependent on the accuracy of the implementation of such vias. In [38], a second-order dual-band filter based on coupled lines loaded with a pair of SIRs and open stubs was presented showing a wide CM rejection bandwidth and high selectivity in the narrow DM passbands. However, the introduction of SIRs increased the total size. Another approach based on stub loaded resonators (SLR) was presented in [39] showing a simple and compact design, easily adaptable to two or three passbands, that presents a good level of CM noise rejection. Nevertheless, the isolation between bands is relatively poor. In [40], a via-free composite right/left-handed (CRLH) resonator was presented showing an excellent CM suppression behavior, both in terms of bandwidth and rejection level. Nonetheless, the bandwidth of the DM passbands is too small for many applications.

In [30], single-and–dual-band stub bandpass filters are presented whose input–output terminals are loaded with resistively terminated *Kth*-order bandstop filters. Thanks to resistive loading, common-mode is neither transmitted nor reflected, but absorbed in the resistors. This technique leads to good results for both DM and CM, but the prize to pay is the complexity in the design and the use of lumped resistors, which are not always desirable at high frequency applications.

The use of magnetic coupling to design balanced single-band BPFs was proposed in [26] as an alternative to the most commonly employed electric coupling. It was demonstrated that magnetic coupling provides inherent strong common-mode rejection when compared with electric coupling. For the simple case of two coupled open-loop resonators, the electric coupling occurs when the resonators are placed in such a way that the gaps of the resonators are placed face-to-face. The boundary conditions under differential- and common-mode excitations barely affect the charge distribution around the open ends of the loops, in such a way that the coupling level for both excitation modes is expected to be similar. By contrast, in the case of magnetically coupled resonators, the open sides are in the opposite orientation. In such case the coupling between resonators can be assimilated to the coupling between two microstrip transmission lines [26]. When common-mode excitation is applied, the symmetry plane of the open loop becomes a virtual open circuit (vanishing currents at that plane) and the coupling mechanism among adjacent resonators remains electrical in nature. On the contrary, a short-circuit condition appears at the same point when differential-mode excitation is applied. Currents around that point are important and the coupling mechanism between adjacent resonators becomes magnetic in nature. For a required coupling level of the differential-mode signal, which is imposed by the filter design specifications, magnetic coupling leads to larger separation between resonators when compared with the electric coupling case [41]. This automatically results into the weakening of the transmission of the common mode signal. Magnetic coupling has also been used for the design of dual-band BPF [42] by using embedded resonators printed on a monolayer structure in order to achieve dual-band operation. Once again, it was proven that magnetic coupling provided an ideal mechanism to reject the common-mode noise. However, the main drawback of the design presented in [42] was the independent control of the features of the two passbands, since both bands were dependent from each other. This is a fact inherent to the use of embedded resonators. Furthermore, in [43,44] the use of magnetic coupling has been extended to the design of balanced diplexers with excellent results that confirm the benefits and flexibility offered by magnetic coupling to design different kinds of differential components. The contribution made in [42] has motivated the authors to present in this paper a different strategy to design balanced dual-band BPFs with more flexibility and independence in what concerns the tuning the two differential passbands. As a case example, the proposed technique is used to design a filter with two differential bands corresponding to two commonly used wireless communications standards (GPS and Wi-Fi bands). It will be shown how, by using a multilayer structure, it is possible to create two differential passbands that can be independently tuned. Two magnetically coupled resonators on the top layer create the upper differential passband, while the lower differential passband is originated from another pair of resonators etched in the middle layer. These resonators are also magnetically coupled. Top layer resonators are excited by means of a edge gap capacitance between the feeding lines and the resonators. Middle layer resonators are excited thanks to the broadside capacitance created between the top and middle layers. Very good DM and CM responses can be achieved using this method, which is experimentally demonstrated for a filter based on two pairs of magnetically coupled open-loop resonators. The most relevant features of the proposed design are: (i) the two differential passbands are almost totally independent due to the asynchronous nature of the resonators and the possibility of independent control of the resonators excitation, (ii) band-to-band isolation about 36 dB can be easily achieved, and (iii) a good CM rejection level, which is inherent to magnetic coupling, can be also easily obtained (more specifically, 39/33 dB measured in the lower/upper DM passbands, respectively).

2. Proposed Structure: Analysis and Design Methodology

2.1. Analysis of the Structure

A 3-D view of the proposed multilayer structure for balanced dual-band operation is depicted in Figure 1. The two dimensional layout for each layer is depicted in Figure 2. The pattern etched on the top layer was used in [26] for the design of a single-band balanced BPF. It basically consists of a pair of open-loop resonators that are magnetically coupled (under DM operation). The middle layer also consists of a pair of open-loop resonators that are magnetically coupled when operating in DM. The difference between the two coupled resonant structures lies on the type of excitation and the values of the resonance frequencies. Thus, top layer resonators are excited via the gap capacitance associated with the gap width s_2^t, while the middle structure is excited by means of the broadside capacitance created between the top feeding lines and the middle layer conducting strips. Figure 3 illustrates the overlapping section between layers which originates the broadside capacitance (indicated by the shaded region). By setting the value of l_{over}, it is possible to control the excitation level of the middle-layer resonators, which is critical to implement the response of the differential passband created by this pair of coupled resonators. Figure 4 plots the simulated electromagnetic response (simulations have been carried out using *ADS Momentum* from Keysight, stationed at Santa Rosa, California, United States of America, USA) of the entire multilayer structure for both DM and CM excitations. Two differential passbands with good CM rejection can be appreciated in such figure, demonstrating the expected behavior of the proposed structure. The design process of the multilayer balanced dual-band bandpass filter that leads to such response is described in detail in the next subsections.

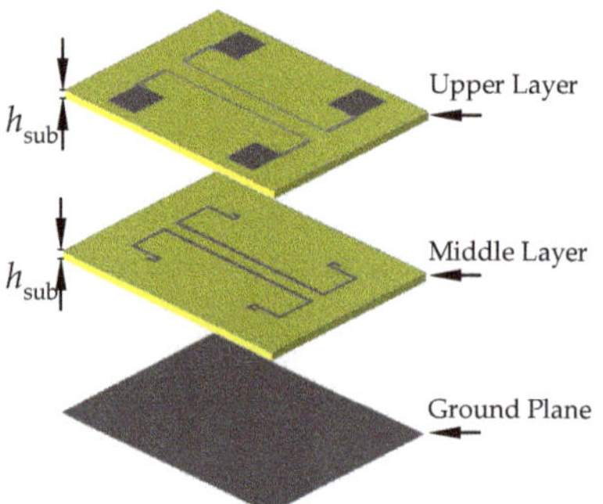

Figure 1. Deployed 3-D view of the proposed planar multilayer balanced dual-band bandpass filter.

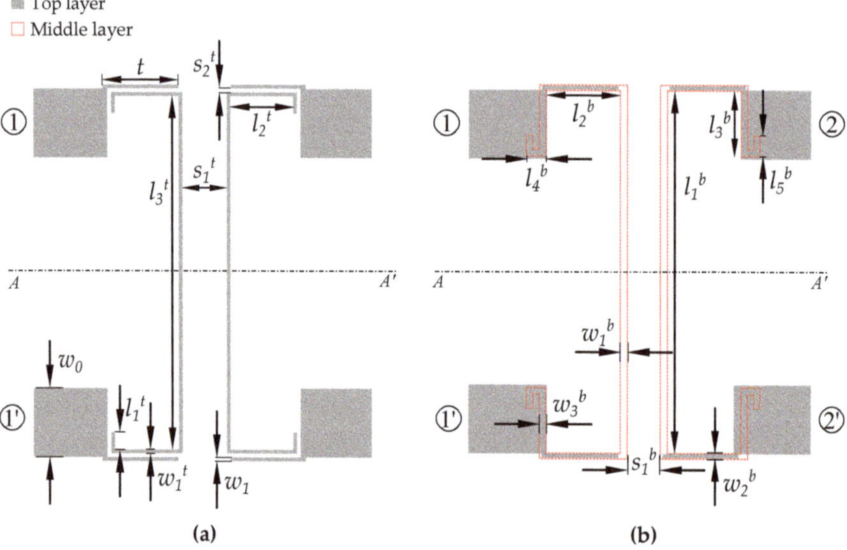

Figure 2. 2-D view of the two printed planes of the structure (not to scale). (**a**) Top layer; (**b**) Middle layer.

Figure 3. Detailed view of the broadside capacitance involved in the feeding of middle layer resonators.

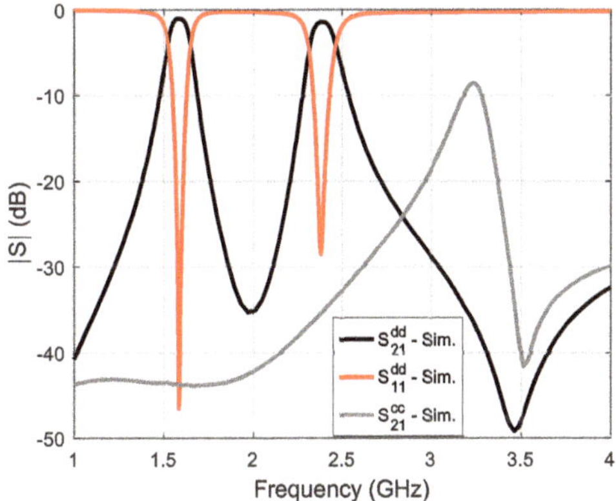

Figure 4. Simulated response of the proposed multilayered balanced dual band bandpass filter. The dimensions (in mm) used for the structure in Figure 2 that give rise to this response are: (i) top layer: $t = 5.8$, $w_0 = 5.1$, $w_1 = 0.2$, $w_1^t = 0.2$, $l_1^t = 1.3$, $l_2^t = 5.05$, $l_3^t = 26.6$, $s_1^t = 0.2$; (ii) middle layer: $w_1^b = 0.55$, $w_2^b = 0.4$, $w_3^b = 0.5$, $l_1^b = 27.2$, $l_2^b = 5.7$, $l_3^b = 4.95$, $l_4^b = 1.5$, $l_5^b = 1.575$, $s_1^b = 2$.

2.2. Design Methodology

When designing a dual-band filter, it is very important that the two passbands can be independently tuned in terms of both center frequency and fractional bandwidth (FBW). In the case at hand, the first point that should be noted is that the middle layer of the structure introduces the lower passband, centered at $f_{01}^d = 1.575$ GHz, the L1 GPS signal frequency, while the top layer resonates at the first Wi-Fi band frequency, $f_{02}^d = 2.45$ GHz. This means that the upper frequency band should be allocated by setting the top layer parameters while lower frequency band must depend, exclusively, on the middle layer geometry. Let's start by studying the resonance frequency of resonators in both layers. In Figure 5, the differential return loss, $|S_{11}^{dd}|$, have been represented when a single couple of top and middle resonators are weakly excited. The dips in the return loss correspond to the resonance frequencies of both resonators. Several parameters from each resonator were varied to show the dependence of those resonance frequencies with the resonator's length. From the simulated results (Figure 5) the following conclusions can be inferred:

1. If middle layer resonators length is kept constant, changing top layer resonators length only affects the upper band center frequency, with very slight effect on the lower band center frequency.
2. If top layer resonators length is kept constant, changing middle layer resonators length only affects the lower band center frequency with negligible effect on the upper band center frequency.

From these results, it is clear that there is flexibility in locating the center frequencies of each passband as long as both frequencies are not very close to each other. In this latter case, resonators on both layers would interact significantly, thus leading to band-to-band dependence breaking the asynchronous behavior between the pair of resonators. This situation would not be desirable for designing purposes. Furthermore, the independence of the passbands is also related to the excitation mechanism of each resonator pair. As mentioned before, top layer resonators are mainly excited due to the electric field confined within the gap of width s_2^t (gap capacitance) while middle layer resonators are excited by means of the electric field concentrated between the feeding lines and the resonator conductor strips (broadside capacitance). This fact, together with the sufficient separation of the two center frequencies, leads to the desired independence exhibited when locating the center frequencies of the differential passbands.

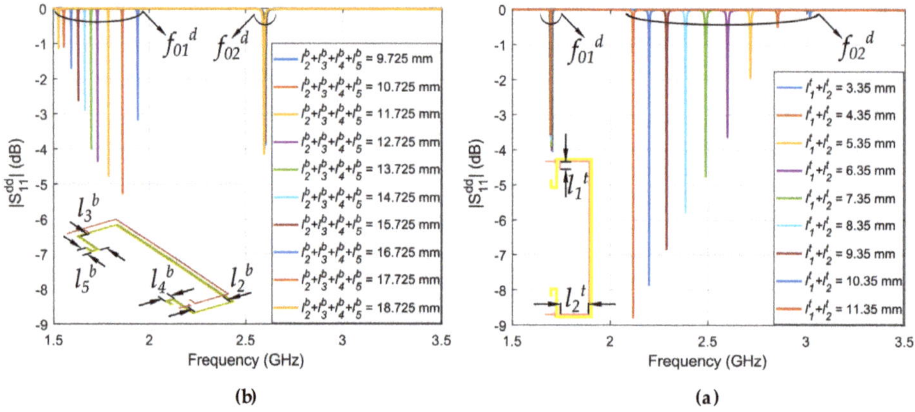

Figure 5. Differential excitation return loss ($|S_{11}^{dd}|$) of the structure composed by a couple of top and middle resonators like those depicted in the insets when they are weakly excited: (**a**) $l_2^b + l_3^b + l_4^b + l_5^b$ is varied; (**b**) $l_1^t + l_2^t$ is varied. In both figures, the remaining physical parameters (see Figure 2) have been kept constant.

Once the independent control of the resonance frequencies has been demonstrated, the next electrical parameter to be considered in the design process is the correct excitation of the resonators to generate two properly matched differential passbands. In this regard, as mentioned above, the parameters that control top and middle layer resonators excitation are mainly s_2^t and l_{over}, respectively. In order to investigate on the influence of such parameters on the matching level of both differential passbands, i.e., $|S_{11}^{dd}|$, Figure 6 shows the differential return loss of the two differential passbands as a function of s_2^t and l_{over}. From this figure it is clear that l_{over} mainly influences the reflection coefficient of the lower differential passband while s_2^t controls the reflection coefficient of the upper differential passband. Note that s_2^t slightly modifies the value of the upper center frequency. However, this problem can be solved by barely modifying some of the other dimensions of the top layers resonators in a very simple manner, as explained in [26]. Finally, the last step in the design process is to verify that the fractional bandwidth of the two differential passbands can be set independently. If this is the case, we have two totally independent differential passbands, which represents the ideal case when designing filters with dual-band operation.

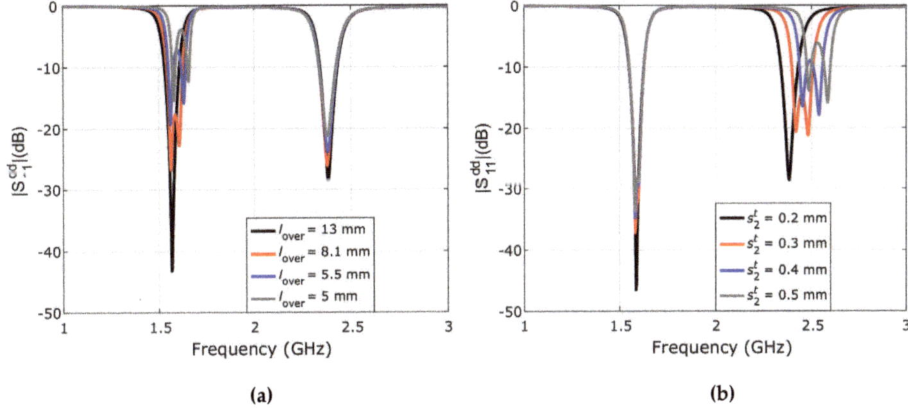

Figure 6. $|S_{11}^{dd}|$ for different values of (**a**) l_{over} and (**b**) s_2^t.

Fractional bandwidth is related to the coupling level between resonators. The closer the resonators are the larger the fractional bandwidth is. Referring to Figure 2, the relevant geometrical parameter that controls the coupling between resonators for the ones located on the top layer is s_1^t, whereas s_1^b is the responsible for the coupling level between resonators in the middle layer. To verify this, fractional bandwidth as a function of s_1^t and s_1^b are shown in Figure 7a,b, respectively. From Figure 7, it is clear that, as it happened with the center frequencies and the differential return loss, the fractional bandwidth of the two differential passbands are also independent from each other. This fact confers the designer full possibilities to adapt the design to other frequencies or other bandwidths that might be necessary for other requirements and specifications.

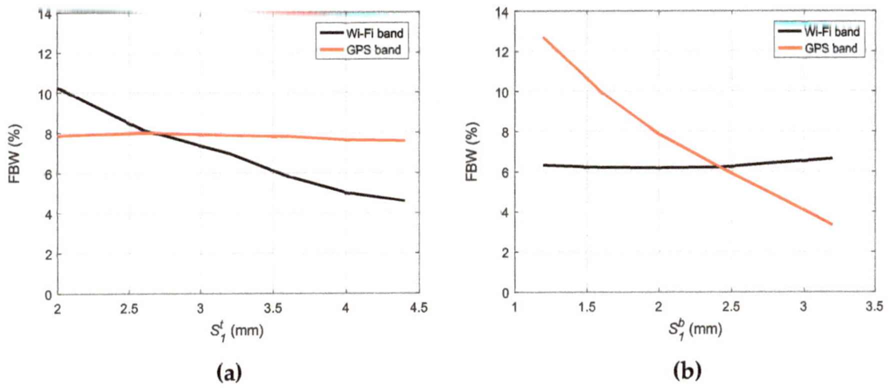

Figure 7. FBW with as a function of the resonators gap: (**a**) s_1^t is adjusted; (**b**) s_1^b is adjusted.

In order to clarify the design process of the balanced dual-band bandpass filter proposed in this paper, the following summary is given below:

1. First, resonator dimensions must be obtained in order to fit their resonance frequencies to the desired passband center frequencies. Since these specific resonators are half-wavelength open loop resonators, their resonance frequencies are mainly determined by their lengths, which can be easily obtained by using well known equations [45] or with the help of the electromagnetic simulator.
2. Next, the fractional bandwidth of each differential passband is set to the desired one by properly adjusting the values of s_1^t and s_1^b.

3.   Finally, the matching level is adjusted by setting the values of s_2^t and l_{over}.

The proposed filter is finally designed and ready to be fabricated and measured.

3. Results

3.1. Prototype Example

In this section, an example of differential dual-band BPF filter design is reported. The first passband corresponds to the L1 GPS signal at 1.575 GHz. A fractional bandwidth of 8.86% is considered. The second passband is designed to be allocated in the first Wi-Fi frequency at 2.45 GHz, with a fractional bandwidth of 5.69%. The filter is implemented on a substrate with the following features: dielectric constant $\varepsilon_{r1} = \varepsilon_{r2} = 3.0$, and thicknesses $h_{sub} = 1.016$ mm. As explained in detail in previous sections, the first step in the design process is to obtain the resonator dimensions to set the two desired center frequencies, which are controlled by resonator lengths (recall that resonance arises when the length of the resonators is half the guided wavelength at the desired center frequencies). Once resonators are designed, design curves in Figures 6 and 7 are used to set the desired matching level and fractional bandwidth. Following this procedure, the final dimensions of the proposed dual-band differential filter in Figure 2 are those of the caption of Figure 4, where we have represented the final simulated filter response. This figure exhibits two differential passbands well allocated. The common-mode rejection level is very satisfactory too. To verify experimentally these results, a prototype has been fabricated and measured. Results are presented in next subsection.

3.2. Experimental Results

In this section, a comparison between the results of the electromagnetic simulations and the measurements of the manufactured prototype is carried out. Figure 8 shows a photograph of the fabricated prototype, which has been fabricated using a mechanical milling machine (LPKF S103 from LPKF Laser & Electronics, stationed at Garbsen, Germany). Measurements have been carried out using Agilent PNA-E8363B with a Test-Set N4420B extension (thus, with a 4-port system), both devices from Keysight, under normal conditions of pressure, temperature and humidity. This network analyzer has a single source, so the DM and CM have been obtained using the program *Physical Layer Test System* from Keysight that allows the user load calibrations and implement a transformation between the actually measured 4×4 scattering matrix of the 4-port single-ended circuit and the 2×2 differential-differential, common-common and differential-common 2×2 scattering matrices of the balanced 2-port system. Simulated and measured DM and CM responses are depicted in Figure 9. This figure shows a good agreement between simulated and measured results, although the measured lower band is slightly shifted when compared with the simulated one. This fact can be explained if we take into account that multilayer structures can suffer from the presence of small air gaps that may appear if layers are not perfectly stuck. Furthermore, the mechanical milling machine might sometimes remove a thin layer of the substrate, and this affects the effective dielectric constant of the substrate. This effect can be minimized using some suitable prepreg layer, which should be included in the simulations, of course. In order to verify the effect of this hypothesized air gap, the response of the differential filter has been simulated considering a 0.025 mm air gap between the stacked dielectric substrates. This response is depicted in Figure 9b. The results show a good agreement between simulations and measurements. The use of a low permittivity substrate has the puspose of reducing the air gap effect, since the closer the dielectric constant is to that of air, the less the effective permittivity will be affected. The use of low permittivity substrates will also decrease the capacitance between layers so that the cross coupling between resonators will also be lower. The measured results show a CM rejection level greater than 39 dB and 33 dB for the first and second DM passbands, respectively. Measured insertion loss at the first and second DM center frequencies is 0.9 dB and 1.9 dB respectively. In addition, band-to-band isolation is better than 36 dB, which is satisfactory for many practical applications.

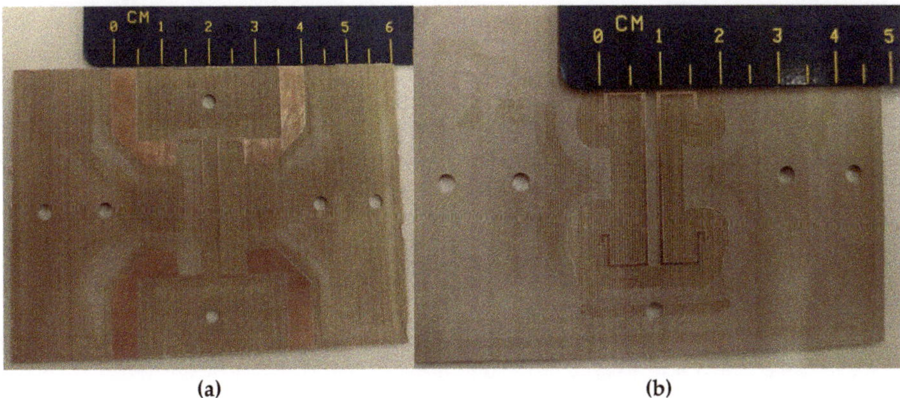

Figure 8. Photograph of the manufactured prototype: (**a**) Top layer; (**b**) middle layer.

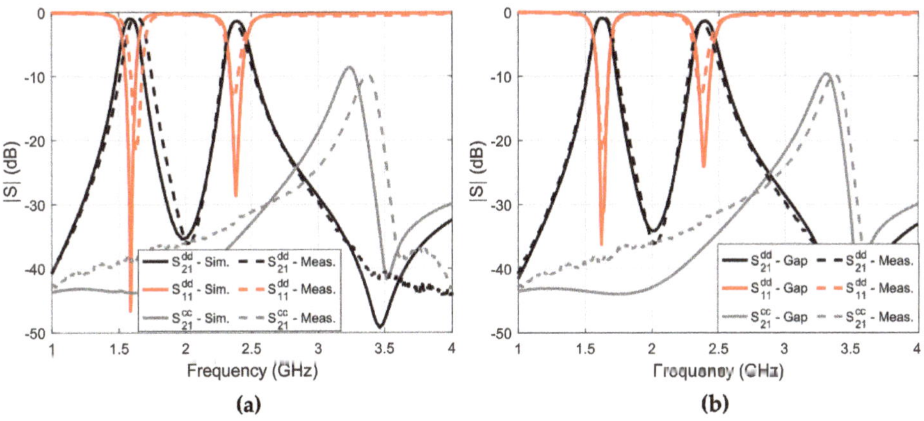

Figure 9. Simulated and measured response of the proposed dual-band bandpass filter: (**a**) without gap; (**b**) with an air gap of 0.025 mm.

4. Discussion

In order to illustrate the benefits of the proposed structure, a comparison between the proposal in this paper and a number of contributions available in the specialized literature is provided in Table 1.

Table 1. Comparison with reported balanced dual-band bandpass filters.

Ref.	Size ($\lambda_g \times \lambda_g$)*	Differential-Mode					Common-Mode
		f_0^d (GHz)	3-dB FBW (%)	IL @f_0^d(dB)	B-to-B IL > 30 dB	CMRR @f_0^d(dB)	$\lvert S_{21}^{cc} \rvert$ > 30dB
[27]	0.187 × 0.285	1.84/2.45	9.2/8.2	2.2/2.6	2–2.3	20.5/20	<30 dB
[28]	0.314 × 0.413	2.46/5.56	18.7/9.5	0.96/1.9	3.05–5	50/37.5	Up to 6.75 GHz
[29]	N/A	2.44/5.19	18/8.7	1.14/2.05	3.6–4.1	41.26/40.35	Up to 7 GHz
[30]	0.275 × 0.13	2.82/3.21	5.2/5.1	1.9/1.7	<30 dB	13.6/7.8	3.2–3.24 GHz
[31]	0.261 × 0.173	2.5/5.27	11/3.98	1.46/2.22	3.4/4.8	38.5/22.9	(2–4.5)/(5.3–7.3) GHz
[32]	0.153 × 0.268	2.5/5.6	5.1/4.8	1.29/1.97	3.05–4.8	34.7/24.1	2.3–2.7 GHz
[33]	0.259 × 0.592	2.44/5.57	16.4/8.62	1.78/2.53	3.15–4.7	36.2/31.1	(1–5.75)/(6.2–8) GHz
[34]	0.417 × 0.377	2.44/5.25	8.61/4.57	2.4/2.82	1.8–3.8	52.6/42.2	Up to 7 GHz
[35]	N/A	1.9/2.8	2.1/3.9	2.05/2.65	<30 dB	40/35.2	Up to 3.2 GHz
[36]	0.5 × 0.2	2.4/3.57	8.33/5.6	0.87/1.9	2.9–3.28	24/29.1	(1.5–2.32)/(2.45–2.9)/ (3–3.6) GHz
[37]	1.23 × 1.23	3.5/5.24	3.14/3.82	1.52/1.65	3.7–4.75	53.48/48.35	Up to 6 GHz
[38]	0.67 × 0.32	0.9/2.49	3.6/2.1	2.67/4.65	1.1–2.3	27.3/35.4	Up to 4.3 GHz
[39]	0.15 × 0.37	2.5/5.8	12.9/4.5	0.77/1.56	<30 dB	41.23/36.44	(1–3)–(5.8–6.8) GHz
[40]	0.131 × 0.161	2.38/3.59	1.33/2.13	1.34/1.03	2.46–3.49	48.6/49	Up to 6 GHz
[42]	0.137 × 0.173	2.45/5.6	7.35/10.39	2.78/2.85	3–4.45	47.2/30.2	(1–3.81)/(5.22–7) GHz
This work	0.237 × 0.148	1.64/2.37	8.86/5.69	0.92/1.94	1.93–2.13	38/31.1	Up to 2.62 GHz

* Guided wavelength (λ_g) at lowest frequency of operation (f_0^d).

In terms of the different relevant parameters, several conclusions can be inferred from the comparison between our design and those reported in the literature (Table 1):

1. **Filter size**. In terms of electrical size, the proposed filter is very competitive. Indeed, is one of the smallest in the table (0.035 λ_g^2, λ_g being the guided wavelength at the center of the lower transmission band, when the average size for those presented in Table 1 is 0.184 λ_g^2). There are only two designs smaller than the one in this contribution [40,42], but it should be taken into account that one of the main advantages of our filter lies on its simplicity. Size reduction through optimization of the geometry of the resonators would be possible but has not been attempted.

2. **Fractional bandwidth**. Our design confers the designer total and independent control on the bandwidth of the two passbands. For our specifications and requirements, the filter perfectly covers both the GPS band and the Wi-Fi band, so there are no issues on this regard. If we look at the results obtained in the considered literature, the average fractional bandwidth is 8.74% (first band) and 5.72% (second band). Both values are very close to those reported in this work. These values can still be increased a little further by adjusting the values of the separation between resonators, as it can be seen in Figure 7 in the paper. Note that using practicable values of such separations, our design would be able to compete with those with the highest values of FBW reported in Table 1.

3. **Insertion loss**. The average IL of the designs presented in Table 1 are 1.63 dB and 2.25 dB for the first and second passbands, respectively. Note that the measured IL values for our design is, once again, competitive with the state-of-the-art, providing one the lowest values of IL for the first passband.

4. **Band-to-band isolation**. The best way of obtaining good band-to-band isolation is to introduce one or more transmission zeros between the passbands. This is not the case in the presented design. Nevertheless, the IL in the frequency region between passbands reaches 36 dB, this being a very good value for many applications. Ten of the designs included in Table 1 exhibit transmission zeroes between passband. Thus, this point could be considered the main drawback of the proposed filter topology and adding elements to introduce transmission zeros would result into a significant improvement, provided the other relevant features of the filter are not significantly affected.

5. **Common-mode Rejection Ratio (CMRR)**. The average CMRR values of the filters in Table 1 for the first and second differential passbands is 37.95 dB and 32.55 dB, respectively. These values are very close to the ones achieved with the proposed design. The advantage of our proposal is that this high level of CMRR is inherent to the coupling scheme and no additional components have to be added to enhance the CM suppression (such as lumped inductors, capacitors or resistors, additional CM rejection stages, additional open-circuit stubs, etc.). The introduction of these elements usually entails disadvantages (increase in complexity and the need of reaching a trade-off to avoid the deterioration of other filter parameters).

In brief, the design presented in this work is one of the smallest reported in the literature, it has low complexity and allows for easy configuration of the two passbands (both, center frequency and FBW). At the same time, leads to high CMRR values without relying on additional elements that might disrupt the differential-mode response. The main deawback of our proposal could be the lack of transmission zeroes in the frequency region between passbands. Nevertheless, the band-to-band isolation level presented in this work is satisfactory for most applications.

5. Conclusions

In this paper, a new dual-band balanced bandpass filter (DB-B-BPF) with inherent common-mode rejection is presented. The proposed structure is based on the use of two pairs of magnetically coupled open-loop resonators implemented in multilayer technology. The lower differential-mode (DM) passband is created by the middle layer resonators, while the upper DM passband is associated with the top layer resonators. By using this configuration, it is possible to design the two DM passbands with independent control of bandwidth and matching level. At the same time, very good band-to-band isolation and common-mode rejection is achieved. To demonstrate the usefulness of this proposal, an example of a filter operating for GPS and Wi-Fi applications (1.58 GHz and 2.4 GHz) was designed, fabricated and measured. Experiments show a Common-Mode Rejection Ratio (CMRR) better than 38 dB and 31.1 dB for the lower and upper DM passbands, respectively. Measured DM insertion loss for both passbands is 0.92/1.94 dB and the obtained 3-dB bandwidth is 8.86/5.69 %.

Author Contributions: J.L.M.d.R., A.L. and A.F.-P. designed, simulated, manufactured and measured the filter prototype. F.M., J.M. and A.J.M.-R. supervised the whole study. All the authors participated in writing the paper and in revising the article. All authors have read and agreed to the published version of the manuscript.

Funding: This work has been funded by the Spanish AEI Ministry of Science, Innovation and Universities and EU Feder Funds (project TEC2017-84724-P). J.M.R. acknowledges financial support of a research scholarship of the Spanish Ministry of Education (reference PRE2018-085677).

Conflicts of Interest: The authors declare no conflict of interest

References

1. Eisenstant, W.R.; Stengel, B.; Thompson, B.M. *Microwave Differential Circuit Design Using Mixed-Mode S-Parameters*; Artech House: Boston, MA, USA, 2006.
2. Martin, F.; Zhu, L.; Hong, J.; Medina, F. *Balanced Microwave Filters*; John Wiley & Sons: New York, NY, USA, 2018.
3. Xia, B.; Wu, L.S.; Mao, J.F. A new balanced-to-balanced power divider/combiner. *IEEE Trans. Microw. Theory Tech.* **2012**, *60*, 2791–2798. [CrossRef]
4. Xia, B.; Wu, L.S.; Ren, S.W.; Mao, J.F. A balanced-to-balanced power divider with arbitrary power division. *IEEE Trans. Microw. Theory Tech.* **2013**, *61*, 2831–2840. [CrossRef]
5. Wu, L.S.; Guo, Y.X.; Mao, J.F. Balanced-to-balanced Gysel power divider with bandpass filtering response. *IEEE Trans. Microw. Theory Tech.* **2013**, *61*, 4052–4062. [CrossRef]
6. Feng, W.; Zhu, H.; Che, W.; Xue, Q. Wideband in-phase and out-of-phase balanced power dividing and combining networks. *IEEE Trans. Microw. Theory Tech.* **2014**, *62*, 1192–1202. [CrossRef]
7. Zhou, Y.; Deng, H.W.; Zhao, Y. Compact balanced-to-balanced microstrip diplexer with high isolation and common-mode suppression. *IEEE Microw. Wirel. Compon. Lett.* **2014**, *24*, 143–145. [CrossRef]
8. Hsiao, C.Y.; Wu, T.L. A novel dual-function circuit combining high-speed differential equalizer and common-mode filter with an additional zero. *IEEE Microw. Wirel. Compon. Lett.* **2014**, *24*, 617–619. [CrossRef]
9. Naqui, J.; Fernández-Prieto, A.; Durán-Sindreu, M.; Mesa, F.; Martel, J.; Medina, F.; Martín, F. Common mode suppression in microstrip differential lines by means of complementary split ring resonators: Theory and applications. *IEEE Trans. Microw. Theory Tech.* **2012**, *60*, 3023–3034. [CrossRef]
10. Olvera-Cervantes, J.L.; Corona-Chávez, A. Microstrip balanced bandpass filter with compact size, extended-stopband and common-mode noise suppression. *IEEE Microw. Wirel. Compon. Lett.* **2013**, *23*, 530–532. [CrossRef]
11. Horestani, A.K.; Durán-Sindreu, M.; Naqui, J.; Fumeaux, C.; Martín, F. S-shaped complementary split ring resonators and their application to compact differential bandpass filters with common-mode suppression. *IEEE Microw. Wirel. Compon. Lett.* **2014**, *24*, 149–151. [CrossRef]
12. Wu, X.H.; Chu, Q.X.; Qiu, L.L. Differential wideband bandpass filter with high-selectivity and common-mode suppression. *IEEE Microw. Wirel. Compon. Lett.* **2013**, *23*, 644–646. [CrossRef]
13. Lin, L.; Bao, J.; Du, J.J.; Wang, Y.M. Differential wideband bandpass filters with enhanced common-mode suppression using internal coupling techniquen. *IEEE Microw. Wirel. Compon. Lett.* **2014**, *24*, 300–302.
14. Xu, X.; Wang, J.; Zhu, L. A new approach to design differential-mode bandpass filters on SIW structure. *IEEE Microw. Wirel. Compon. Lett.* **2013**, *23*, 635–637. [CrossRef]
15. Feng, W.; Che, W. Novel wideband differential bandpass filters based on T-shaped structure. *IEEE Trans. Microw. Theory Tech.* **2012**, *60*, 1560–1568. [CrossRef]

16. Wu, X.H.; Chu, Q.X. Compact differential ultra-wideband band-pass filter with common-mode suppression. *IEEE Microw. Wirel. Compon. Lett.* **2012**, *22*, 456–458. [CrossRef]
17. Lu, Y.J.; Chen, S.Y.; Hsu, P. A differential-mode wideband bandpass filter with enhanced common-mode suppression using slotline resonator. *IEEE Microw. Wirel. Compon. Lett.* **2012**, *22*, 503–505. [CrossRef]
18. Vélez, P.; Naqui, J.; Fernández-Prieto, A.; Durán-Sindreu, M.; Bonache, J.; Martel, J.; Medina, F.; Martín, F. Differential bandpass filter with common-mode suppression based on open split ring resonators and open complementary split ring resonators. *IEEE Microw. Wirel. Compon. Lett.* **2013**, *23*, 22–24. [CrossRef]
19. Feng, W.; Che, W.; Ma, Y.; Xue, Q. Compact wideband differential bandpass filters using half-wavelength ring resonator. *IEEE Microw. Wirel. Compon. Lett.* **2013**, *23*, 81–83. [CrossRef]
20. Shi, J.; Shao, C.; Chen, J.X.; Lu, Q.Y.; Peng, Y.; Bao, Z.H. Compact low-loss wideband differential bandpass filter with high common-mode suppression. *IEEE Microw. Wirel. Compon. Lett.* **2013**, *23*, 480–482. [CrossRef]
21. Wu, C.H.; Wang, C.H.; Chen, C.H. Novel balanced coupled-line bandpass filters with common-mode noise suppression. *IEEE Trans. Microw. Theory Tech.* **2007**, *55*, 287–295. [CrossRef]
22. Wu, C.H.; Wang, C.H.; Chen, C.H. Stopband-extended balanced bandpass filter using coupled stepped-impedance resonators. *IEEE Microw. Wirel. Compon. Lett.* **2007**, *17*, 507–509. [CrossRef]
23. Lin, S.C.; Yeh, C.Y. Stopband-extended balanced filters using both $\lambda/4$ and $\lambda/2$ SIRs with common mode suppression and improved passband selectivity. *Prog. Electromagn. Res.* **2012**, *128*, 215–228. [CrossRef]
24. Shi, J.; Chen, J.; Tang, H.; Zhou, L. Differential bandpass filter with high common-mode rejection ratio inside the differential-mode passband using controllable common-mode transmission zero. In Proceedings of the 2013 IEEE International Wireless Symposium (IWS), Beijing, China, 14–18 April 2013; doi:10.1109/IEEE-IWS.2013.6616742. [CrossRef]
25. Wang, H.; Tam, K.W.; Ho, S.K.; Kang, W.; Wu, W. Short-ended self-coupled ring resonator and its application for balanced filter design. *IEEE Microw. Wirel. Compon. Lett.* **2014**, *24*, 312–314. [CrossRef]
26. Fernández-Prieto, A.; Lujambio, A.; Martel, J.; Medina, F.; Mesa, F.; Boix, R. Simple and Compact Balanced Bandpass Filters Based on Magnetically Coupled Resonators. *IEEE Trans. Microw. Theory Tech.* **2015**, *63*, 1843–1853. [CrossRef]
27. Shi, J.; Xue, Q. Balanced Bandpass Filters Using Center-Loaded Half-Wavelength Resonators. *IEEE Trans. Microw. Theory Tech.* **2010**, *58*, 970–977.
28. Shi, J.; Xue, Q. Dual-Band and Wide-Stopband Single-Band Balanced Bandpass Filter With High Selectivity and Common-Mode Suppression. *IEEE Trans. Microw. Theory Tech.* **2010**, *58*, 2204–2212. [CrossRef]
29. Guo, X.; Zhu, L.; Wu, W. Balanced Wideband/Dual-Band BPFs on a Hybrid Multimode Resonator With Intrinsic Common-Mode Rejection. *IEEE Trans. Microw. Theory Tech.* **2016**, *64*, 1997–2005. [CrossRef]
30. Gómez-García, R.; Muñoz-Ferreras, J.-M.; Feng, W.; Psychogiou, D. Balanced Symmetrical Quasi-Reflectionless Single-and Dual-Band Bandpass Planar Filters *IEEE Microw. Wirel. Compon. Lett.* **2018**, *28*, 798–800. [CrossRef]
31. Fernández-Prieto, A.; Martel, J.; Medina, F.; Mesa, F.; Qian, S.; Hong, J.; Naqui, J.; Martín, F. Dual-Band Differential Filter using Broadband Common-Mode Rejection Artificial Transmission Line. *Prog. Electromagn. Res.* **2013**, *139*, 779–797. [CrossRef]
32. Bagci, F.; Fernández-Prieto, A.; Lujambio, A.; Martel, J.; Bernal, J.; Medina, F. Compact Balanced Dual-Band Bandpass Filter Based on Modified Coupled-Embedded Resonators. *IEEE Microw. Wirel. Compon. Lett.* **2017**, *27*, 31–33. [CrossRef]
33. Shi, J.; Xue, Q. Novel Balanced Dual-Band Bandpass Filter Using Coupled Stepped Impedance-Resonators. *IEEE Microw. Wirel. Compon. Lett.* **2010**, *20*, 19–21.
34. Lee, C.H.; Hsu, C.I.G.; Hsu, C.C. Balanced Dual-Band BPF with Stub-Loaded SIRs for Common-Mode Suppression. *IEEE Microw. Wirel. Compon. Lett.* **2010**, *20*, 70–72. [CrossRef]
35. Wang, K.; Zhu, L.; Wong, S.W.; Chen, D.; Guo, Z.C. Balanced dual-band BPF with intrinsic common-mode suppression on double-layer substrate. *Electron. Lett.* **2015**, *51*, 705–707. [CrossRef]
36. Cho, Y.H.; Yun, S.W. Design of Balanced Dual-Band Bandpass Filters Using Asymmetrical Coupled Lines. *IEEE Trans. Microw. Theory Tech.* **2013**, *61*, 2814–2820. [CrossRef]
37. Li, P.; Chu, H.; Zhao, D.; Chen, R.S. Compact Dual-Band Balanced SIW Bandpass Filter With Improved Common-Mode Suppression. *IEEE Microw. Wirel. Compon. Lett.* **2017**, *27*, 347–349. [CrossRef]

Appl. Sci. **2020**, *10*, 3113

38. Yang, L.; Choi, W.W.; Tam, K.W.; Zhu, L. Balanced Dual-Band Bandpass Filter With Multiple Transmission Zeros Using Doubly Short-Ended Resonator Coupled Line. *IEEE Microw. Wirel. Compon. Lett.* **2015**, *63*, 2225–2232. [CrossRef]

39. Wei, F.; Guo, Y.J.; Qin, P.Y.; Shi, X.W. Compact Balanced Dual- and Tri-band Bandpass Filters Based on Stub Loaded Resonators. *IEEE Microw. Wirel. Compon. Lett.* **2015**, *25*, 76–78. [CrossRef]

40. Song, Y.; Liu, H.; Zhao, W.; Wen, P.; Wang, Z. Compact Balanced Dual-Band Bandpass Filter With High Common-Mode Suppression Using Planar Via-Free CRLH Resonator. *IEEE Microw. Wirel. Compon. Lett.* **2018**, *28*, 996–998. [CrossRef]

41. Hong, J.S. *Microstrip Filters for RF/Microwave Applications*; John Wiley & Sons: New York, NY, USA, 2011.

42. Fernández-Prieto, A.; Martel, J.; Ugarte-Parrado, P.J.; Lujambio, A.; Martínez-Ros, A.J.; Martín, F.; Medina, F.; Boix, R.R. Compact balanced dual-band bandpass filter with magnetically coupled embedded resonators. *IET Microw. Antennas Propag.* **2019**, *13*, 492–497. [CrossRef]

43. Fernández-Prieto, A.; Lujambio, A.; Martel, J.; Medina, F.; Martín, F.; Boix, R.R. Compact Balanced-to-Balanced Diplexer Based on Split-Ring Resonators Balanced Bandpass Filters. *IEEE Microw. Wirel. Compon. Lett.* **2018**, *28*, 218–220. [CrossRef]

44. Fernández-Prieto, A.; Lujambio, A.; Martel, J.; Medina, F.; Martín, F.; Boix, R.R. Balanced-to-Balanced Microstrip Diplexer Based on Magnetically Coupled Resonators. *IEEE Access* **2018**, *6*, 18536–18547. [CrossRef]

45. Makimoto, M.; Yamahista, S. *Microwave Resonators and Filters for Wireless Communication. Theory, Design and Application*; Springer: New York, NY, USA, 2001.

applied sciences

Article

High Refractive Index Electromagnetic Devices in Printed Technology Based on Glide-Symmetric Periodic Structures

Philip Arnberg *, Oscar Barreira Petersson, Oskar Zetterstrom, Fatemeh Ghasemifard and Oscar Quevedo-Teruel

Division for Electromagnetic Engineering, KTH Royal Institute of Technology, 114-28 Stockholm, Sweden; opete@kth.se (O.B.P.); oskarz@kth.se (O.Z.); fatemehg@kth.se (F.G.); oscarqt@kth.se (O.Q.-T.)
* Correspondence: parnberg@kth.se

Received: 12 March 2020; Accepted: 21 April 2020; Published: 5 May 2020

Abstract: We demonstrate the beneficial effects of introducing glide symmetry in a two-dimensional periodic structure. Specifically, we investigate dielectric parallel plate waveguides periodically loaded with Jerusalem cross slots in three configurations: conventional, mirror- and glide-symmetric. Out of these three configurations, it is demonstrated that the glide-symmetric structure is the least dispersive and has the most isotropic response. Furthermore, the glide-symmetric structure provides the highest effective refractive index, which enables the realization of a broader range of electromagnetic devices. To illustrate the potential of this glide-symmetric unit cell, a Maxwell fish-eye lens is designed to operate at 5 GHz. The lens is manufactured in printed circuit board technology. Simulations and measurements are in good agreement and a measured peak transmission coefficient of -0.5 dB is achieved.

Keywords: glide symmetry; higher symmetries; Maxwell fish-eye lens; metasurface; periodic structures; printed circuit board

1. Introduction

A periodic structure is said to possess a higher symmetry if it has an additional geometrical symmetry beyond its translational symmetry. One particular type of higher symmetry is glide symmetry. A glide-symmetric structure is invariant under a translation by half a period and a reflection in a glide plane [1]. Higher symmetries in one dimensional periodic structures were extensively studied for electromagnetic purposes in the 1960s and 1970s [1–5] (some earlier works on helical twist-symmetric structures exist [6,7]). These works outlined a theoretical basis for the analysis of higher-symmetric structures but did not highlight all details of the propagation characteristics in these structures. Recently, it was demonstrated that higher-symmetric periodic structures provide attractive properties; for instance, the possibility of realizing a higher value of the effective refractive index [8–11] and a reduction of the intrinsic frequency dispersion in periodic structures [12–17]. It was also discovered that glide-symmetric metasurfaces can support edge modes [18], provide an increased control of the anisotropy [19] and effective permeability [20] in two-dimensional (2D) periodic structures. In other words, the recent studies demonstrate that higher symmetries provide an additional degree of freedom to control the wave propagation in periodic structures [11,14,19,21–25], which enables the design of novel millimeter wave devices; such as lenses [26–28], filters [29,30], phase shifters [31–33], polarizers [34–37] and low-cost efficient high-frequency waveguides and antennas [38–41]. Additionally, glide-symmetric holey structures have been used to suppress the leakage between waveguide flanges, which enables contact-less measurements of electromagnetic devices [42]. The attractive properties found in higher-symmetric structures have inspired several works on numerical methods for the analysis [16,43–50]. These methods also give valuable physical insight.

Due to the forthcoming increase of the operation frequency to millimeter waves in communication systems [51], the requirements on transmitters and receivers are increasing. For instance, highly directive antennas are required to compensate for the increase in the free space path loss [52]. Graded index lenses are a viable option to achieve highly directive antennas at millimeter wave frequencies [52–59]. One example of a graded index lens is the Maxwell fish-eye (MFE) lens. The MFE lens is a rotationally symmetric lens that converges electromagnetic waves passing through a point in space to its antipodal point [60], and it is an attractive solution for imaging systems. Due to the rotational symmetry, the focusing property of the MFE lens is maintained for all angles and there is an ongoing debate among scientists whether or not the MFE lens has the ability of perfect imaging [61–71]. Furthermore, the half MFE (HMFE) lens has received significant attention [53–58] due to its ability to transform a spherical wave into a planar wave, similar to a Luneburg lens [26,59,72]. Contrary to the Luneburg lens, reflection occurs at output of the HMFE lens. Additionally, since the rotational symmetry is broken in the HMFE lens, the scanning capabilities are limited. However, the HMFE lens is half the size of the Luneburg lens and is thus useful for size-constrained applications where only a limited scanning is required.

In this work, we study the effect of introducing glide symmetry into a 2D periodic structure. Specifically, we analyze a parallel plate waveguiding structure where one or both of the conductors are loaded with slots. It is demonstrated that a less dispersive and higher effective refractive index is obtained in a glide-symmetric structure, compared to its non-glide-symmetric counterparts. The large achievable range of effective refractive indices enables a variety of microwave devices to be realized. Here, we design a planar MFE lens in printed circuit board (PCB) technology using the analyzed glide-symmetric structure. In fact, without glide symmetry, the performance of the designed lens is severely limited. A prototype of the lens is constructed in order to corroborate the simulations.

2. Glide Symmetry

A periodic structure is glide-symmetric if it is invariant under a translation and a reflection. A 2D glide-symmetric structure implemented in a parallel plate waveguide (PPW) is exemplified in Figure 1. The inset displays a top view of the unit cell. The displacement between the discontinuities in the top and bottom plates is $p/2$, where p is the periodicity.

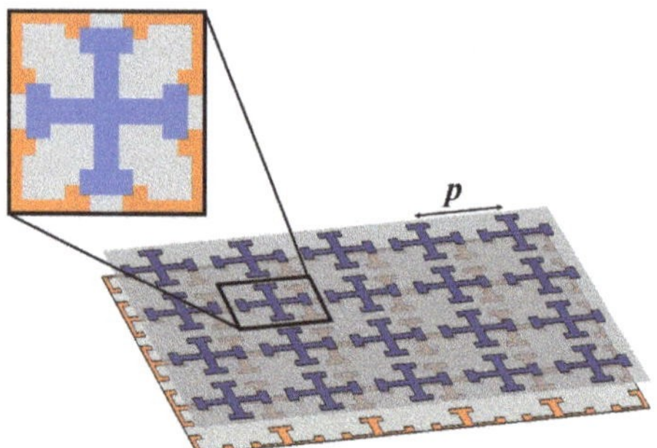

Figure 1. Illustration of a 2D glide-symmetric structure implemented in a PPW. The orange and blue crosses are placed on the bottom and top conductor, respectively. The inset displays a top view of the unit cell.

The effect of the displacement between the discontinuities on the propagation characteristics in a 2D periodic structure is illustrated in Figure 2, where the dispersion diagram is obtained using the eigenmode solver of CST Microwave Studio [73]. The analyzed structure is a PPW with Jerusalem cross slots placed in both conductors, as illustrated in Figure 2a. The displacement of the slots, s, is varied from perfectly aligned ($s = 0$) to glide-symmetric ($s = p/2$). For a shift smaller than $p/2$, there is a stop band between the first and second modes and the modes experience significant dispersion near this stop band. However, for $s = p/2$, the stop band is suppressed and the modes connect. Therefore, the group velocity is no longer required to be zero at the band edge and the dispersion is reduced. Furthermore, it is observed that the effective refractive index in the glide-symmetric configuration is increased, compared to the non-glide-symmetric structures. In the following sections, these properties of glide-symmetric structures are employed to design a planar Maxwell fish-eye lens.

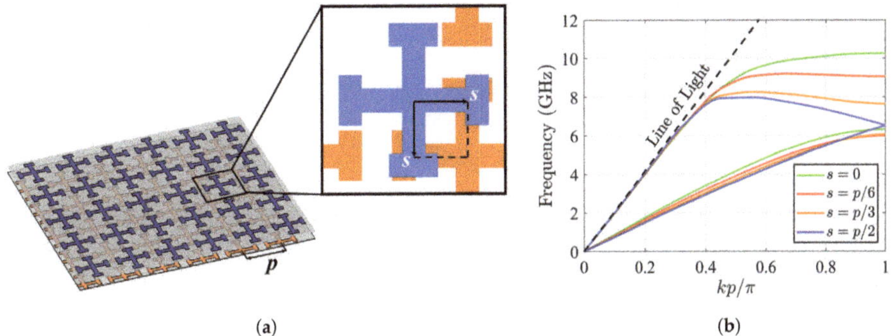

(a)   (b)

Figure 2. Illustration of the effect of displacing the cross in the two PPW layers: (**a**) periodic structure with a top view of the unit cell as an inset, and (**b**) simulated dispersion diagram for different displacements.

3. Lens Design

3.1. Unit Cell Design

The studied unit cells are composed of two metallic layers with a dielectric in between, illustrated in Figure 3a. A Jerusalem cross slot is placed in the top metallic layer, as illustrated in Figure 3b. The three studied structures have different bottom conductors. Two reference structures are analyzed, with a solid bottom conductor (Figure 3c) and with a mirror-symmetric Jerusalem cross slot in the bottom conductor (Figure 3d). The third structure has a glide-symmetric Jerusalem cross slot in the bottom conductor (Figure 3e). The metallic layers have a thickness of 0.035 mm.

The three unit cells are modeled and simulated with the Eigenmode solver of CST. The normalized effective refractive index (with $n_0 = \sqrt{\varepsilon_r}$) for a parametric sweep of the different structures is presented in Figure 3f–h. The parameters are (unless otherwise specified) $L_e = 0.5$ mm, $L_t = 2.9$ mm, $w_b = 1.3$ mm, $w_c = 0.5$ mm, $p = 7.2$ mm, $h = 0.8$ mm, and $\varepsilon_r = 2.53$. From the parametric sweep, we observe that the glide-symmetric structure has the least dispersive response and provides a higher effective refractive index. Furthermore, in Figure 3h, the effective refractive index for waves propagating in two directions (0° and 45° with respect to the x-axis) is presented. The response of the glide-symmetric structure is almost isotropic, which is not the case over a wider range of frequencies compared to the reference structures. A less dispersive behavior entails a broader bandwidth for the device. A relative measure of the dispersion can be obtained by comparing the change of the effective refractive index around a reference value at a given frequency. For instance, if we allow a maximum change of 2 % around the normalized effective refractive index value 2 at 5 GHz the bandwidth is 4.7 %, 5.2 % and 19 % for the conventional, mirror- and glide-symmetric structures. Hence, the 2 %-deviation bandwidth around an effective refractive index of 2 is almost four times larger in the glide-symmetric structure, compared to its non-glide counterparts. The acceptable deviation from the nominal value, and the bandwidth increase, depends on the intended application.

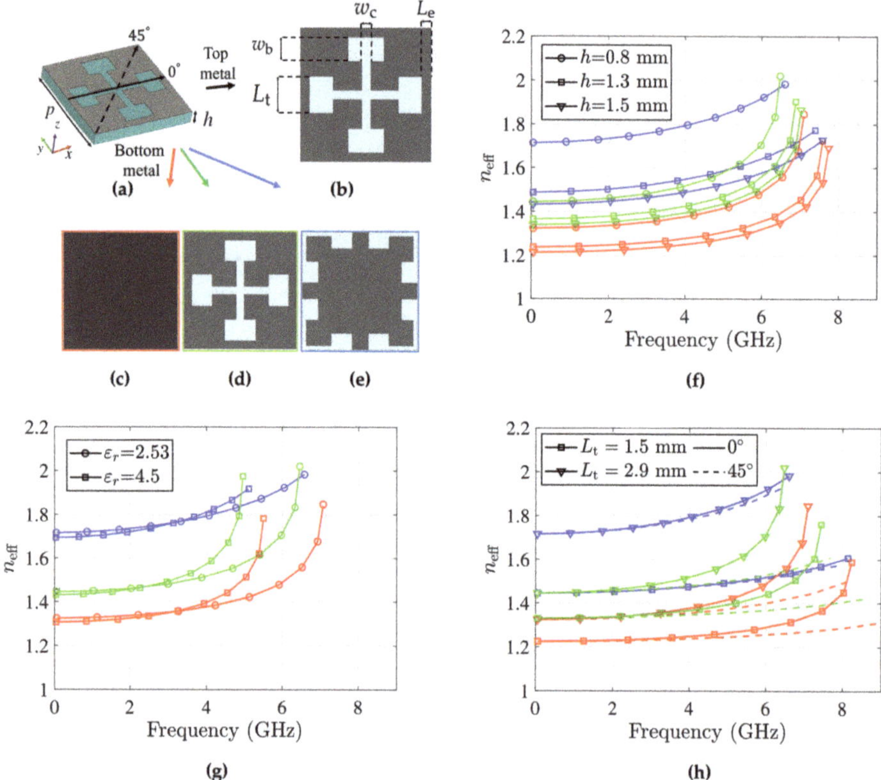

Figure 3. Dispersion analysis for the dielectric parallel plate waveguide loaded with Jerusalem cross slots. The unit cell is illustrated in (**a**) perspective view, and (**b**) top view. Three unit cell configurations with different bottom metallic conductors are studied: (**c**) conventional, (**d**) mirror-, and (**e**) glide-symmetric. Effective refractive index for different values of: (**f**) the substrate thickness, h, (**g**) the substrate permittivity, ε_r, and (**h**) the parameter L_t. The level of isotropy is illustrated in (**h**). The red, green and blue curves correspond to the conventional, mirror- and glide-symmetric unit cells. The parameters are (unless otherwise specified) $L_e = 0.5$ mm, $L_t = 2.9$ mm, $w_b = 1.3$ mm, $w_c = 0.5$ mm, $p = 7.2$ mm, $h = 0.8$ mm, and $\varepsilon_r = 2.53$. A normalization factor of $\sqrt{\varepsilon_r}$ is applied to n_{eff}.

3.2. Maxwell Fish-Eye Lens

The refractive index profile of an MFE lens is given by

$$n(r) = \frac{2n_0}{1 + (\frac{r}{a})^2} \tag{1}$$

where n_0 is the refractive index of the surrounding media, a is the radius of the lens and r is the radial position in the lens. The refractive index must range from n_0 to $2n_0$. Such variation can be achieved with the glide-symmetric structure by varying the parameters L_t and L_e, as illustrated in Figure 4. The refractive index profile (1) is realized by spatially varying L_t and L_e throughout the lens. A rendition of the lens is presented in Figure 5a. The radius of the lens, a, is 130 mm. The prototype of the lens is displayed in Figure 5b and it is manufactured on a $h = 0.8$ mm thick Teflon substrate ($\varepsilon_r = 2.53$ and $\tan \delta = 0.001$).

Metallic vias are placed along the perimeter of the lens to emulate a cylindrical metallic wall. In this way, the feed and image point can be moved inside the lens [74]. This is done to avoid uncontrolled reflection at the termination of the PPW, which facilitates the characterization of the lens. The vias have a diameter of 0.3 mm, are placed at the radius a, and are separated by 0.5 mm (center to center).

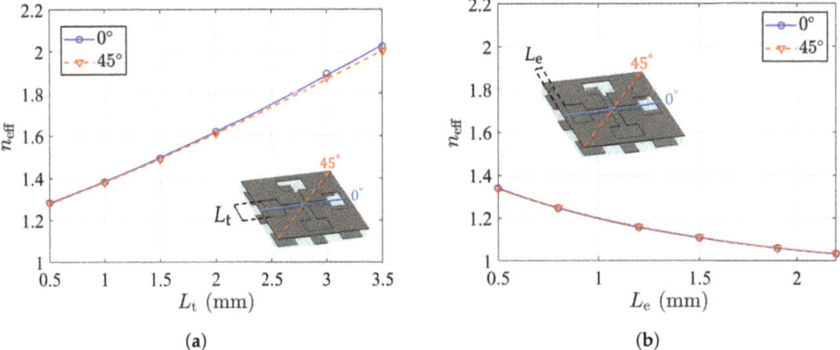

Figure 4. Effective refractive index at 5 GHz for a parametric sweep of: (**a**) L_t and (**b**) L_e. The blue and red line represent a wave traveling in the 0° and 45° directions, with respect to the x-axis. The effective refractive index is normalized with $\sqrt{\varepsilon_r} = n_0$. The parameters are $L_e = 0.5$ mm (Figure 4a), $L_t = 1$ mm (Figure 4b), $w_b = 1$ mm, $w_c = 0.5$ mm (Figure 4a), $w_c = 1$ mm (Figure 4b), $p = 7.2$ mm, $h = 0.8$ mm, and $\varepsilon_r = 2.53$.

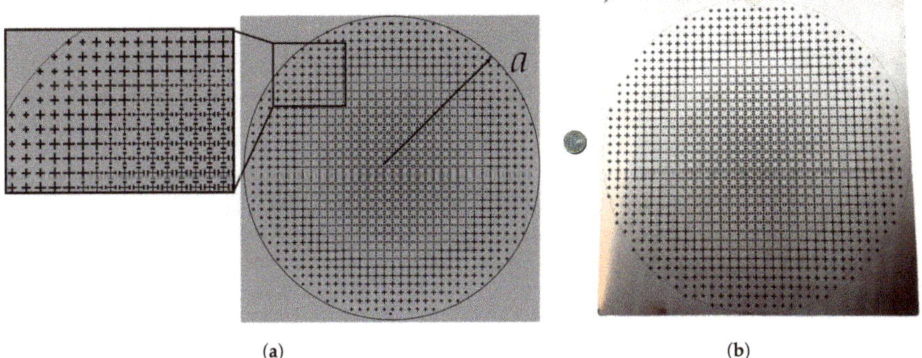

Figure 5. Top view of the designed lens: (**a**) rendition, and (**b**) manufactured prototype.

3.3. Feed Design

Due to the metallic shielding of the lens, an omnidirectional feed can be used. Therefore, the lens is fed by a probe placed at the radius $b = 124.5$ mm. A probe is also placed at the corresponding image point. In order to match the impedance of the lens to the generator impedance (50 Ω), a single-stub matching circuit is designed in microstrip technology. The matching circuit is illustrated in Figure 6a. The simulated reflection coefficient for the matching circuit when loaded with a 0.8 mm high PPW is presented in Figure 6b. The manufactured matching circuit is presented in the inset of Figure 6b. The dimensions are $w = 2.27$ mm , $D = 1.0$ mm, $\Delta r = 0.7$ mm, $d = 6.5$ mm, $l = 19.8$ mm and $s_l = 32.9$ mm and the circuit is manufactured on a 0.8 mm thick Teflon substrate ($\varepsilon_r = 2.53$ and $\tan \delta = 0.001$).

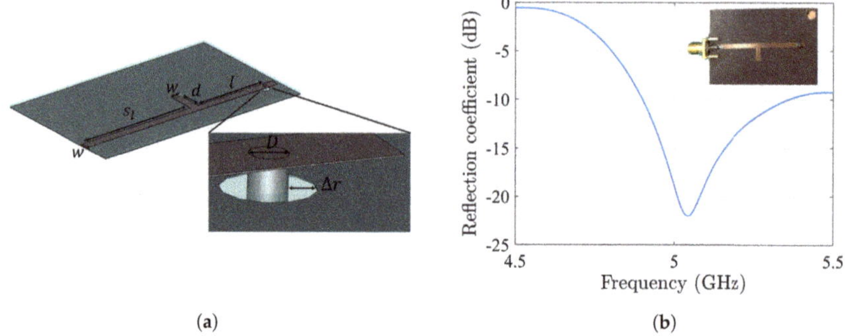

(a)   (b)

Figure 6. Simulation results of the matching circuit: (**a**) 3D model and (**b**) reflection coefficient when loaded with a dielectric PPW. The manufactured prototype is presented in the inset. The dimensions are $w = 2.27\,\text{mm}$, $D = 1.0\,\text{mm}$, $\Delta r = 0.7\,\text{mm}$, $d = 6.5\,\text{mm}$, $l = 19.8\,\text{mm}$ and $s_1 = 32.9\,\text{mm}$. The circuit is manufactured on a 0.8 mm thick Teflon substrate ($\varepsilon_r = 2.53$ and $\tan\delta = 0.001$).

4. Results

The full lens is simulated in the Time domain solver of CST. Waveguide ports are used to excite a quasi-TEM mode on the matching circuits which in turn are connected to the lens with vias. The imaging properties of the lens are illustrated in Figure 7a with the absolute value of the electric field sampled at 4.8 GHz. The lens is excited at probe 1, which is connected to the matching circuit through a hole in the bottom conductor of the PPW. An image is created at the antipodal point, where probe 2 is placed, similarly connected to a matching circuit.

The simulated and measured scattering parameters are presented in Figure 7b. The simulation and measurement agrees well, apart from a slight frequency shift. However, since the distance between peaks in the $|S_{11}|$ is similar in the simulations and measurements, the reflections occur at the same location in the structure. The frequency shift is accounted for by the error margin in the relative permittivity of the substrate. The measured $|S_{21}|$ has a peak value of -0.5 dB and it is above -3 dB from 4.63 to 5.03 GHz. The bandwidth depends on the refractive index variation and the feed design. Two more probes are connected to the lens at the same radius as the feed probe, but displaced 22.5° and 45° from the image point. The measured transmission coefficients to these probes are also included in Figure 7b and are below -25 dB.

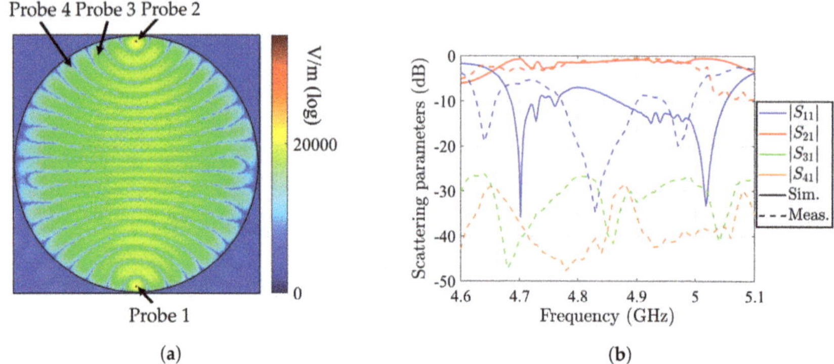

(a)   (b)

Figure 7. (**a**) Absolute value of the electric field distribution at 4.8 GHz obtained with CST simulation. (**b**) Simulated (solid lines) and measured (dashed lines) scattering parameters.

5. Conclusions

In this paper, we have studied the effect of applying glide symmetry to a 2D periodic structure. We have demonstrated that a glide-symmetric structure presents a more isotropic and a less dispersive response compared to conventional periodic structures. Furthermore, a glide-symmetric structure can obtain a higher effective refractive index, which enables a wider range of devices to be realized. These properties have been used to design a PPW MFE lens in PCB technology operating at 5 GHz.

Without applying glide symmetry, the highest refractive index of the MFE lens could not be practically reached for this specific thickness of the slab and dielectric constant. Measurements of the lens agree well with the simulations, and the measured peak transmission coefficient through the lens is -0.5 dB.

Author Contributions: Conceptualization, O.Q.-T.; formal analysis, P.A. and O.B.P.; visualization, P.A. and O.B.P.; writing, original draft preparation, P.A. and O.B.P. ; writing, review and editing, O.Z., F.G. and O.Q.-T.; supervision, O.Z, F.G. and O.Q.-T. All authors have read and agreed to the published version of the manuscript.

Funding: This research was partly funded by the Stiftelsen Åforsk project H-Materials (18-302).

Conflicts of Interest: The authors declare no conflict of interest

References

1. Hessel, A.; Chen, M.H.; Li, R.; Oliner, A. Propagation in periodically loaded waveguides with higher symmetries. *Proc. IEEE* **1973**, *61*, 183–195. [CrossRef]
2. Crepeau, P.; Mcisaac, P. Consequences of symmetry in periodic structures. *Proc. IEEE* **1964**, *52*, 33–43. [CrossRef]
3. Trigg, G.L. Higher Symmetries. *Phys. Rev. Lett.* **1965**, *14*, 479–479.
4. Mittra, R.; Laxpati, S. Propagation in a wave guide with glide reflection symmetry. *Can. J. Phys.* **1965**, *43*, 353–372. [CrossRef]
5. Kieburtz, R.; Impagliazzo, J. Multimode propagation on radiating traveling-wave structures with glide-symmetric excitation. *IEEE Trans. Antennas Propag.* **1970**, *18*, 3–7. [CrossRef]
6. Pocklington, H.C. Electrical oscillations in wires. *Proc. Camb. Philos. Soc.* **1897**. *9*, 324-332.
7. Sensiper, S. Electromagnetic Wave Propagation on Helical Structures (A Review and Survey of Recent Progress). *Proc. IRE* **1955**, *43*, 149–161. [CrossRef]
8. Chang, T.; Kim, J.U.; Kang, S.K.; Kim, H.; Kim, D.K.; Lee, Y.H.; Shin, J. Broadband giant-refractive-index material based on mesoscopic space-filling curves. *Nat. Commun.* **2016**, *7*. [CrossRef] [PubMed]
9. Cavallo, D.; Felita, C. Analytical Formulas for Artificial Dielectrics With Nonaligned Layers. *IEEE Trans. Antennas Propag.* **2017**, *65*, 5303–5311. [CrossRef]
10. Cavallo, D. Dissipation Losses in Artificial Dielectric Layers. *IEEE Trans. Antennas Propag.* **2018**, *66*, 7460–7465. [CrossRef]
11. Dahlberg, Oskar.; Ghasemifard, Fatemeh.; Valerio, Guido.; Quevedo-Teruel, Oscar. Propagation characteristics of periodic structures possessing twist and polar glide symmetries. *EPJ Appl. Metamatetrials* **2019**, *6*, 14. [CrossRef]
12. Mitchell-Thomas, R.C.; Sambles, J.R.; Hibbins, A.P. High index metasurfaces for graded lenses using glide symmetry. In Proceedings of the 2017 11th European Conference on Antennas and Propagation (EUCAP), Paris, France, 19–24 March 2017; pp. 1396–1397.
13. Camacho, M.; Mitchell-Thomas, R.C.; Hibbins, A.P.; Sambles, J.R.; Quevedo-Teruel, O. Designer surface plasmon dispersion on a one-dimensional periodic slot metasurface with glide symmetry. *Opt. Lett.* **2017**, *42*, 3375–3378. [CrossRef] [PubMed]
14. Camacho, M.; Mitchell-Thomas, R.C.; Hibbins, A.P.; Sambles, J.R.; Quevedo-Teruel, O. Mimicking glide symmetry dispersion with coupled slot metasurfaces. *Appl. Phys. Lett.* **2017**, *111*, 121603. [CrossRef]
15. Dahlberg, O.; Mitchell-Thomas, R.; Quevedo-Teruel, O. Reducing the dispersion of periodic structures with twist and polar glide symmetries. *Sci. Rep.* **2017**, *7*, 10136. [CrossRef] [PubMed]

16. Ghasemifard, F.; Norgren, M.; Quevedo-Teruel, O. Dispersion Analysis of 2-D Glide-Symmetric Corrugated Metasurfaces Using Mode-Matching Technique. *IEEE Microw. Wirel. Components Lett.* **2018**, *28*, 1–3. [CrossRef]

17. Ghasemifard, F. Periodic Structures with Higher Symmetries: Analysis and Applications. Doctoral Thesis, Electrical Engineering School of Electrical Engineering and Computer Science, KTH Royal Institute of Technology, Stockholm, Sweden, 2018.

18. de Pineda, J.D.; Hibbins, A.P.; Sambles, J.R. Microwave edge modes on a metasurface with glide symmetry. *Phys. Rev. B* **2018**, *98*, 205426. [CrossRef]

19. Ebrahimpouri, M.; Quevedo-Teruel, O. Ultrawideband Anisotropic Glide-Symmetric Metasurfaces. *IEEE Antennas Wirel. Propag. Lett.* **2019**, *18*, 1547–1551. [CrossRef]

20. Ebrahimpouri, M.; Herran, L.F.; Quevedo-Teruel, O. Wide-Angle Impedance Matching Using Glide-Symmetric Metasurfaces. *IEEE Microw. Wirel. Components Lett.* **2020**, *30*, 8–11. [CrossRef]

21. Quesada, R.; Martín-Cano, D.; García-Vidal, F.J.; Bravo-Abad, J. Deep-subwavelength negative-index waveguiding enabled by coupled conformal surface plasmons. *Opt. Lett.* **2014**, *39*, 2990–2993. [CrossRef]

22. Padilla, P.; Herrán, L.F.; Tamayo-Domínguez, A.; Valenzuela-Valdés, J.F.; Quevedo-Teruel, O. Glide Symmetry to Prevent the Lowest Stopband of Printed Corrugated Transmission Lines. *IEEE Microw. Wirel. Compon. Lett.* **2018**, *28*, 750–752. [CrossRef]

23. Ghasemifard, F.; Norgren, M.; Quevedo-Teruel, O. Twist and Polar Glide Symmetries: An Additional Degree of Freedom to Control the Propagation Characteristics of Periodic Structures. *Sci. Rep.* **2018**, *8*, 11266–11266. [CrossRef] [PubMed]

24. Tamayo-Dominguez, A.; Fernandez-Gonzalez, J.M.; Quevedo-Teruel, O. One-Plane Glide-Symmetric Holey Structures for Stop-Band and Refraction Index Reconfiguration. *Symmetry* **2019**, *11*, 495. [CrossRef]

25. Padilla, P.; Palomares-Caballero, Á.; Alex-Amor, A.; Valenzuela-valdés, J.; Fernández-González, J.M.; Quevedo-Teruel, O. Broken Glide-Symmetric Holey Structures for Bandgap Selection in Gap-Waveguide Technology. *IEEE Microw. Wirel. Components Lett.* **2019**, *29*, 327–329. [CrossRef]

26. Quevedo-Teruel, O.; Ebrahimpouri, M.; Ng Mou Kehn, M. Ultrawideband Metasurface Lenses Based on Off-Shifted Opposite Layers. *IEEE Antennas Wirel. Propag. Lett.* **2016**, *15*, 484–487. [CrossRef]

27. Dahlberg, O.; Valerio, G.; Quevedo-Teruel, O. Fully Metallic Flat Lens Based on Locally Twist-Symmetric Array of Complementary Split-Ring Resonators. *Symmetry* **2019**, *11*, 581. [CrossRef]

28. Shanei, M.M.; Fathi, D.; Ghasemifard, F.; Quevedo-Teruel, O. All-silicon reconfigurable metasurfaces for multifunction and tunable performance at optical frequencies based on glide symmetry. *Sci. Rep.* **2019**, *9*, 13641. [CrossRef]

29. Monje-Real, A.; Fonseca, N.J.G.; Zetterstrom, O.; Pucci, E.; Quevedo-Teruel, O. Holey Glide-Symmetric Filters for 5G at Millimeter-Wave Frequencies. *IEEE Microw. Wirel. Components Lett.* **2020**, *30*, 31–34. [CrossRef]

30. Mouris, B.A.; Fernandez-Prieto, A.; Thobaben, R.; Martel, J.; Mesa, F.; Quevedo-Teruel, O. Increment of the Bandwidth of Mushroom-type EBG Structures with Glide Symmetry. *IEEE Trans. Microw. Theory Tech.* **2020**, *68*, 1365–1375. [CrossRef]

31. Rajo-Iglesias, E.; Ebrahimpouri, M.; Quevedo-Teruel, O. Wideband Phase Shifter in Groove Gap Waveguide Technology Implemented With Glide-Symmetric Holey EBG. *IEEE Microw. Wirel. Components Lett.* **2018**, *28*, 476–478. [CrossRef]

32. Quevedo-Teruel, O.; Dahlberg, O.; Valerio, G. Propagation in Waveguides With Transversal Twist-Symmetric Holey Metallic Plates. *IEEE Microw. Wirel. Components Lett.* **2018**, *28*, 858–860. [CrossRef]

33. Palomares-Caballero, A.; Alex-Amor, A.; Padilla, P.; Luna, F.; Valenzuela-Valdes, J. Compact and Low-Loss V-Band Waveguide Phase Shifter Based on Glide-Symmetric Pin Configuration. *IEEE Access* **2019**, *7*, 31297–31304. [CrossRef]

34. Wei, Z.; Cao, Y.; Fan, Y.; Yu, X.; Li, H. Broadband polarization transformation via enhanced asymmetric transmission through arrays of twisted complementary split-ring resonators. *Appl. Phys. Lett.* **2011**, *99*, 221907. [CrossRef]

35. Zhao, Y.; Belkin, M.; Alù, A. Twisted optical metamaterials for planarized ultrathin broadband circular polarizers. *Nat. Commun.* **2012**, *3*, 870. [CrossRef] [PubMed]

36. Askarpour, A.; Zhao, Y.; Alu, A. Wave propagation in twisted metamaterials. *Phys. Rev. (Condens. Matter Mater. Phys.)* **2014**, *90*, 054305. [CrossRef]

37. Zhao, Y.; Askarpour, A.N.; Sun, L.; Shi, J.; Li, X.; Alù, A. Chirality detection of enantiomers using twisted optical metamaterials. *Nat. Commun.* **2017**, *8*, 14180. [CrossRef] [PubMed]

38. Ebrahimpouri, M.; Rajo-Iglesias, E.; Sipus, Z.; Quevedo-Teruel, O. Cost-Effective Gap Waveguide Technology Based on Glide-Symmetric Holey EBG Structures. *IEEE Trans. Microw. Theory Tech.* **2018**, *66*, 927–934. [CrossRef]

39. Ebrahimpouri, M.; Quevedo-Teruel, O.; Rajo-Iglesias, E. Design Guidelines for Gap Waveguide Technology Based on Glide-Symmetric Holey Structures. *IEEE Microw. Wirel. Components Lett.* **2017**, *27*, 542–544. [CrossRef]

40. Zetterstrom, O.; Pucci, E.; Padilla, P.; Wang, L.; Quevedo-Teruel, O. Low-Dispersive Leaky Wave Antennas for mmWave Point-to-Point High-Throughput Communications. *IEEE Trans. Antennas Propag.* **2020**, *68*, 1322–1331. [CrossRef]

41. Chen, Q.; Zetterstrom, O.; Pucci, E.; Palomares-Caballero, A.; Padilla, P.; Quevedo-Teruel, O. Glide-Symmetric Holey Leaky-Wave Antenna with Low Dispersion for 60-GHz Point-to-Point Communications. *IEEE Trans. Antennas Propag.* **2019**, *68*, 1925–1936. [CrossRef]

42. Ebrahimpouri, M.; Algaba Brazalez, A.; Manholm, L.; Quevedo-Teruel, O. Using Glide-Symmetric Holes to Reduce Leakage Between Waveguide Flanges. *IEEE Microw. Wirel. Components Lett.* **2018**, *28*, 473–475. [CrossRef]

43. Valerio, G.; Sipus, Z.; Grbic, A.; Quevedo-Teruel, O. Accurate Equivalent-Circuit Descriptions of Thin Glide-Symmetric Corrugated Metasurfaces. *IEEE Trans. Antennas Propag.* **2017**, *65*, 2695–2700. [CrossRef]

44. Valerio, G.; Ghasemifard, F.; Sipus, Z.; Quevedo-Teruel, O. Glide-Symmetric All-Metal Holey Metasurfaces for Low-Dispersive Artificial Materials: Modeling and Properties. *IEEE Trans. Microw. Theory Tech.* **2018**, *66*, 3210–3223. [CrossRef]

45. Chen, Q.; Ghasemifard, F.; Valerio, G.; Quevedo-Teruel, O. Modeling and Dispersion Analysis of Coaxial Lines With Higher Symmetries. *IEEE Trans. Microw. Theory Tech.* **2018**, *66*, 4338–4345. [CrossRef]

46. Ghasemifard, F.; Norgren, M.; Quevedo-Teruel, O.; Valerio, G. Analyzing Glide-Symmetric Holey Metasurfaces Using a Generalized Floquet Theorem. *IEEE Access* **2018**, *6*, 71743–71750. [CrossRef]

47. Bagheriasl, M.; Valerio, G. Bloch Analysis of Electromagnetic Waves in Twist-Symmetric Lines. *Symmetry* **2019**, *11*, 620. [CrossRef]

48. Bagheriasl, M.; Quevedo-Teruel, O.; Valerio, G. Bloch Analysis of Artificial Lines and Surfaces Exhibiting Glide Symmetry. *IEEE Trans. Microw. Theory Tech.* **2019**, *67*, 2618–2628. [CrossRef]

49. Sipus, Z.; Bosiljevac, M. Modeling of Glide-Symmetric Dielectric Structures. *Symmetry* **2019**, *11*, 805. [CrossRef]

50. Mesa, F.; Rodríguez-Berral, R.; Medina, F. On the Computation of the Dispersion Diagram of Symmetric One-Dimensionally Periodic Structures. *Symmetry* **2018**, *10*, 307. [CrossRef]

51. Wang, Y.; Li, J.; Huang, L.; Jing, Y.; Georgakopoulos, A.; Demestichas, P. 5G Mobile: Spectrum Broadening to Higher-Frequency Bands to Support High Data Rates. *IEEE Veh. Technol. Mag.* **2014**, *9*, 39–46. [CrossRef]

52. Quevedo-Teruel, O.; Ebrahimpouri, M.; Ghasemifard, F. Lens Antennas for 5G Communications Systems. *IEEE Commun. Mag.* **2018**, *56*, 36–41. [CrossRef]

53. Fuchs, B.; Lafond, O.; Rondineau, S.; Himdi, M.; Le Coq, L. Off-Axis Performances of Half Maxwell Fish-Eye Lens Antennas at 77 GHz. *IEEE Trans. Antennas Propag.* **2007**, *55*, 479–482.

54. Fuchs, B.; Lafond, O.; Palud, S.; Le Coq, L.; Himdi, M.; Buck, M.; Rondineau, S. Comparative Design and Analysis of Luneburg and Half Maxwell Fish-Eye Lens Antennas. *IEEE Trans. Antennas Propag.* **2008**, *56*, 3058–3062.

55. Mei, Z.L.; Bai, J.; Niu, T.M.; Cui, T.J. A Half Maxwell Fish-Eye Lens Antenna Based on Gradient-Index Metamaterials. *IEEE Trans. Antennas Propag.* **2012**, *60*, 398–401.

56. Huang, M.; Yang, S.; Gao, F.; Quarfoth, R.; Sievenpiper, D. A 2-D Multibeam Half Maxwell Fish-Eye Lens Antenna Using High Impedance Surfaces. *IEEE Antennas Wirel. Propag. Lett.* **2014**, *13*, 365–368.

57. Xu, H.X.; Wang, G.M.; Tao, Z.; Cai, T. An Octave-Bandwidth Half Maxwell Fish-Eye Lens Antenna Using Three-Dimensional Gradient-Index Fractal Metamaterials. *IEEE Trans. Antennas Propag.* **2014**, *62*, 4823–4828.

58. Shi, Y.; Li, K.; Wang, J.; Li, L.; Liang, C.H. An Etched Planar Metasurface Half Maxwell Fish-Eye Lens Antenna. *IEEE Trans. Antennas Propag.* **2015**, *63*, 3742–3747.

Appl. Sci. **2020**, *10*, 3216

59. Quevedo-Teruel, O.; Miao, J.; Mattsson, M.; Algaba-Brazalez, A.; Johansson, M.; Manholm, L. Glide-Symmetric Fully Metallic Luneburg Lens for 5G Communications at Ka-Band. *IEEE Antennas Wirel. Propag. Lett.* **2018**, *17*, 1588–1592.

60. Maxwell, J. Solutions of problems. *Camb. Dublin Math. J.* **1854**, *8*, 76.

61. Leonhardt, U. Reply to comment on perfect imaging without negative refraction. *New J. Phys.* **2011**, *13*, 058001.

62. Leonhardt, U.; Philbin, T. Reply to "Comment on 'Perfect imaging with positive refraction in three dimensions'". *Phys. Rev. A* **2010**, *82*, 057802.

63. Miñano, J.C.; Marqués, R.; González, J.C.; Benítez, P.; Delgado, V.; Grabovickic, D.; Freire, M. Super-resolution for a point source better than $\lambda/500$ using positive refraction. *New J. Phys.* **2011**, *13*. doi:10.1088/1367-2630/13/12/125009. [CrossRef]

64. Minano, J.; Sanchez-Dehesa, J.; Gonzalez, J.; Benitez, P.; Grabovickic, D.; Carbonell, J.; Ahmadpanahi, H. Experimental evidence of super-resolution better than [lamada]/105 with positive refraction. *New J. Phys.* **2014**, *16*, 33015. [CrossRef]

65. Tyc, T.; Zhang, X. Perfect lenses in focus. *Nature* **2011**, *480*, 42–43. [CrossRef] [PubMed]

66. Blaikie, R.J. Comment on perfect imaging without negative refraction. *New J. Phys.* **2010**, *12*, 058001. [CrossRef]

67. Merlin, R. Comment on "perfect imaging with positive refraction in three dimensions". *Phys. Rev. A-At. Mol. Opt. Phys.* **2010**, *82*, 057801. [CrossRef]

68. Merlin, R. Maxwell's fish-eye lens and the mirage of perfect imaging. *J. Opt.* **2011**, *13*, 024017. [CrossRef]

69. Blaikie, R.J. Perfect imaging without refraction? *New J. Phys.* **2011**, *13*, 125006. [CrossRef]

70. Kinsler, P.; Favaro, A. Comment on reply to comment on perfect imaging without negative refraction. *New J. Phys.* **2011**, *13*, 028001. [CrossRef]

71. Tyc, T.; Danner, A. Resolution of Maxwell's fisheye with an optimal active drain. *New J. Phys.* **2014**, *16*, 1–11. [CrossRef]

72. Luneberg, K. *Mathematical Theory of Optics*; University of California Press: Berkeley, CA, USA, 1944.

73. CST Microwave Studio, Version 2018. Available online: http://www.cst.com/ (accessed on 21 November 2019).

74. Leonhardt, U. Perfect imaging without negative refraction. *New J. Phys.* **2009**, *11*, 093040. [CrossRef]

Article

CPW-Fed Transparent Antenna for Vehicle Communications

Jorge Iván Trujillo-Flores [1], Richard Torrealba-Meléndez [1,*], Jesús Manuel Muñoz-Pacheco [1], Marco Antonio Vásquez-Agustín [1], Edna Iliana Tamariz-Flores [2], Edgar Colín-Beltrán [3] and Mario López-López [1]

[1] Faculty of Electronics Sciences, Autonomous University of Puebla, Puebla 72000, Mexico; jorge.trujillof@correo.buap.mx (J.I.T.-F.); jesusm.pacheco@correo.buap.mx (J.M.M.-P.); marco.vasqueza@correo.buap.mx (M.A.V.-A.); mario.lopezlop@correo.buap.mx (M.L.-L.)
[2] Faculty of Computer Sciences, Autonomous University of Puebla, Puebla 72000, Mexico; iliana.tamariz@correo.buap.mx
[3] CONACYT—Instituto Nacional de Astrofísica, Óptica y Electrónica, Tonantzintla, Puebla 72840, Mexico; edgarcb@inaoep.mx
*   Correspondence: richard.torrealba@correo.buap.mx; Tel.: +52-(222)-229-5500 (ext. 7400)

Received: 22 July 2020; Accepted: 27 August 2020; Published: 29 August 2020

Abstract: In this paper, a fully transparent multiband antenna for vehicle communications is designed, fabricated, and analyzed. The antenna is coplanar waveguide-fed to facilitate its manufacture and increase its transmittance. An indium-tin-oxide film, a type of transparent conducting oxide, is selected as the conductive material for the radiation path and ground plane, with 8 ohms/square sheet resistance. The substrate is glass with a relative permittivity of 5.5, and the overall dimensions of the optimized design are 50 mm × 17 mm × 1.1 mm. The main antenna parameters, namely, sheet resistance, reflection coefficient, and radiation diagram, were measured and compared with simulations. The proposed antenna fulfills the frequency requirements for vehicular communications according to the IEEE 802.11p standard. Additionally, it covers the frequency bands from 1.82 to 2.5 GHz for possible LTE communications applied to vehicular networks.

Keywords: coplanar waveguides; vehicular networks; IEEE 802.11p; indium-tin oxide (ITO); transparent antenna

---

1. Introduction

Communications have acquired an important role in the development of concepts such as the internet of things (IoT) [1]. The IoT provides emerging and novel applications to transform cities in smart cities by improving the quality and performance of its public services [2,3]. Smart cities are capable of sensing, integrating, and analyzing critical information in city operation and evolution to ensure sustainability and quality of life through the advancement of urban electronic communications in interoperable systems [4–6]. One of the principal goals in smart cities is to count on smart mobility. Smart mobility consists of transport that defines an innovative infrastructure for traffic and transport that saves resources by using new technologies for maximum accessibility and efficiency for citizens [7]. In the IoT field, there exist specific networks that can improve and grow smart mobility; these are known as vehicular networks (VN) [8]. These networks carry out vehicle to vehicle (V2V), vehicle to infrastructure (V2I) [9], and vehicle to everything (V2X) communication [8]. The development and implementation of VN contribute to the integration of mobility with everything. That is, being reached by the concept of smart [1,10], which ranges from smart devices and homes to smart cities. Currently, several communications systems integrate with vehicles due to the growing demand for connectivity in vehicular environments [11,12]. Such integration is regulated under the framework of the IEEE 802.11p standard [11,13]. At present, several works on VN have employed communications systems based on IEEE 802.11p—this standard is the core for the wireless access for vehicular environments [14–17]. The IEEE 802.11p standard allows vehicular communications defined in the physical (PHY) and medium access control (MAC) layers in the dedicated short-range communications (DSRC) protocol stack [18]. In DSRC, the 5.85–5.925 GHz range is divided into seven channels of 10-MHZ bandwidth each, which is the half bandwidth of 802.11a [18]. This standard offers short transmission distance less than 300 m and data rates ranging from 6 to 27 Mb/s [14]. Besides, 802.11p uses a carrier frequency of 5.9 GHz [16]. Moreover, recent developments have been carried out proposing a system that combines the traditional 802.11p standard VANET networks with LTE networks to form a hybrid Cloud-VANET to provide an efficient routing protocol with low latency and high reliability along with less congestion [19]. Reference [20] presents an analysis of the obtained results of a V2V framework developed on VANETs to reduce large amounts of traffic from LTE networks.

A relevant element in the design of vehicular communication systems is the antenna, which must satisfy the appropriate radiation specifications [21]. Besides, the antennas must ensure that there is no intervention with the vehicle operation. Usually, antennas are made of metal on dielectric substrates, which can be unattractive for the growing need to optimize spaces and offer better technical and aesthetic performance. Some examples of this are in references [22,23]—present antennas for 802.11p standard, but are implemented in an opaque substrate. Even in reference [22], the proposed design is a volumetric structure. Furthermore, in [24], a fractal antenna design is presented—this antenna was fabricated using Polyvinyl chloride as a transparent substrate, but it uses copper for the radiation element, making it not completely transparent. Therefore, demand is emerging for research, development, and use of transparent antennas that can be integrated into windshields and windows, solar cell panels, mobile devices, and in the automotive industry, [25,26]. Finally, it has to be mentioned that there exist other transparent antenna designs for vehicular communications, but they operate within the UHF and VHF bands [27,28].

Additionally, transparent antennas start to be an option in the application of RF energy harvesting. Some examples of these applications are described by [29,30]. For the development of transparent antennas, various materials and configurations have been tested to improve efficiency. Some of the transparent conductive films that have been tested for use in this type of antennas include silver-coated polyester film (AgHT-8) [31], indium-tin-oxide (ITO) [32], Indium-zinc-tin oxide (IZTO) [33], Silver grid layer coating (AgGL) [34], and zinc oxide heavily doped with gallium (GZO) [25]. Besides, some transparent antennas designs employ mesh metal films printed on glass substrates—these metal layers present a low sheet resistance [35,36]. Recently, mesh metal films are applied in transparent antennas at 60 GHz to improve their performance [37].

On the other hand, there are many investigations of transparent antenna geometries. In reference [26] a transparent antenna for WLAN application in 5 GHz is introduced, whose topology is inspired in a flower with six petals built into indium-tin oxide film (ITO) as a transparent radiation element. Several antennas designs use coplanar wave feed [31,38,39], because the transparent conductive film is on just one side of the substrate. In [31], a coplanar waveguide-fed transparent antenna for extended ultrawideband applications is presented, using AgHT-8 thin film as conductive material with a staircase-shaped rectangular radiator and a modified partial ground. Moreover, references [38,39] report a multiband transparent antenna using a coplanar waveguide for 5G applications.

Notwithstanding the vast developed area, a remaining problem for designing transparent antennas based on conductive films with sheet resistance over 5 Ohms, (mainly with ITO) [26,29,30], is to obtain a multiband behavior with narrow bandwidths. Typically, the transparent antennas with ITO show a fractional bandwidth higher than 20% generating a UWB response. In this work, a multiband operation achieves, and the frequencies bands are well defined.

In this work, the design of a fully transparent antenna using ITO-coated glass with a coplanar waveguide (CPW)-fed slot dipole with a multiband operation, IEEE 8002.11p (5.9 GHz), and LTE sub 6 GHz bands, is proposed for future vehicular communications applications. In order to obtain the desired frequencies, a modified ground plane was introduced with symmetrical rectangular stubs [40,41]. The antenna characterization is performed by considering the measured reflection coefficient and the radiation pattern. Even though the ITO has a high sheet resistance, the experimental results obtained satisfy the frequency requirements for the standard IEEE 802.11 and LTE sub 6 GHz bands. Besides, the experimental results are similar in behavior to the simulated ones, but with a slight shift in frequency. Nevertheless, this good concordance is unusual in other reported transparent antennas based on conductive films with a high sheet impedance. The main contributions of the proposed design are a fully transparent antenna that operates at frequencies for vehicular communications standards, and small size.

2. Materials and Methods

In this research, Adafruit commercial indium-tin-oxide (ITO)-coated glass with a thickness of 1850 Å was employed to fabricate the proposed antenna. The transparent conductive oxide (TCO) sheet resistance (Rs) was estimated by the well-known method of the four-point probe [42] performed over thirteen positions on a film sample, as can be seen in Figure 1. Table 1 presents those results and the final average for Rs of 7.943 ohm/square that was used for the design and simulations of the antenna. The substrate is glass with relative permittivity and thickness of 5.5 and 1.1 mm, respectively. In the development of antennas, those factors are significant to determine the frequency responses and antenna efficiency [32]. Therefore, it is important to incorporate them into the antenna design and simulation process. Moreover, a coplanar waveguide-fed slot antenna with impedance matching stub was chosen to obtain a multiband response through a crossmatch technique for a proper impedance matching [40,41], and allowing both the radiator and the ground plane to be on the same side for fabrication ease. High-frequency structural simulator (HFSS, ANSYS, PA, US) was used to design and optimize the chosen topology for the transparent antenna through parametric analysis.

Figure 1. Schematic positions on indium-tin oxide (ITO) film placed on a glass slide for sheet resistance (Rs)/square measurement.

Table 1. Rs/Square Measurements Obtained on ITO Film.

Position	Resistance (Ω)
1	7.831
2	7.909
3	7.970
4	8.008
5	8.035
6	7.997
7	7.914
8	7.954
9	7.914
10	7.900
11	7.922
12	7.958
13	7.950
Average	7.943

Through a parametric analysis in the HFSS simulator, the optimized antenna dimensions were obtained, and they are shown in Figure 2 and described in Table 2, respectively. The antenna feedline was calculated to obtain a coplanar waveguide (CPW) with a characteristic impedance of 50 ohms. The proposed transparent antenna is a capacitively coupled CPW-fed slot, dipole antenna with two pairs of protruded slot lines with different lengths, W1 and W2. The resonant frequencies correspond to the two different resonant paths. The two resonance frequencies for the design are 1.9 GHz and 5.9 GHz, which are used in LTE and 802.11p, respectively. Besides, in the parametric analysis, we made variations over the parameters W1, W2, and Wp1, to reach the desired frequency bands performing the matching impedance and observe their impact on the antenna design.

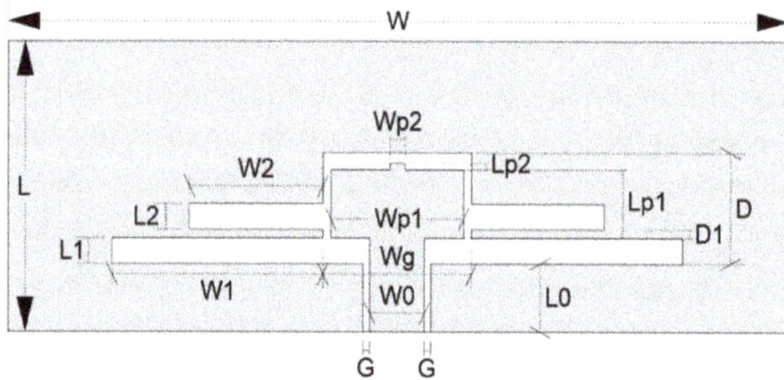

Figure 2. Antenna configuration with dimensions.

Table 2. Dimensions of the Proposed Antenna.

Parameter	Size (mm)	Parameter	Size (mm)
W	50	L	17
Wg	9.5	Lp1	4
Wp1	8.5	Lp2	0.4
Wp2	1	L0	4
W0	3.5	L1	1.5
W1	13.5	L2	1.5
W2	8.5	D	6.5
G	0.4	D1	0.5

In [40], it is mentioned that the resonant frequencies are independently controlled by a common aperture and the three pairs of respective thin slots. However, in our design, the variation of the length W1 shift both frequencies, as shown in Figure 3a. Moreover, Figure 3b shows variations in the higher frequency due to change in the length W2, while the lower frequency is fixed. Similar behavior is observed in Figure 3c when varying the width of the central patch, the high frequency is mainly affected, whereas the lower band is minimally changed. The reason is attributable to the fact that lengths of W2 and Wp1 are closer to the shortest wavelength [41]. It is important to mention that in our simulations, the SMA connector was considered.

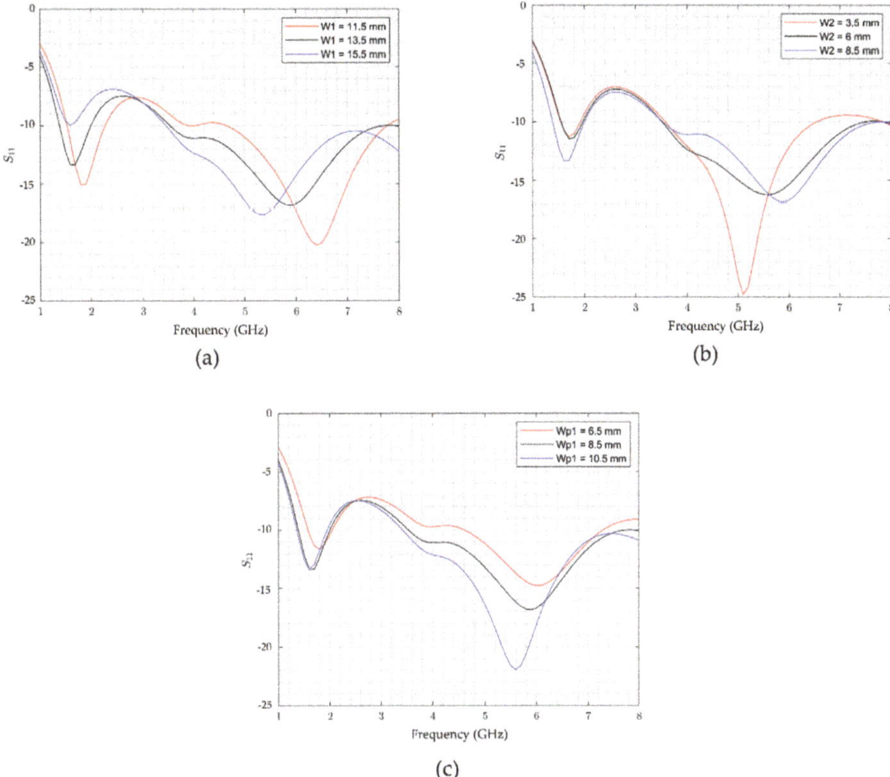

Figure 3. Simulated reflection coefficient of the ITO antenna for three parameters variations: (a) W1; (b) W2; (c) Wp1.

3. Results and Discussion

After obtaining the results through numerical simulation, the antenna was manufactured, and measurements of return loss, gain, and radiation diagrams were performed. The transparent antenna was fabricated by using a photolithography process. The transfer of the antenna geometry to the ITO-coated glass was made using a photosensitive film, which was exposed to UV light through a mask with the antenna geometry. Then the area exposed was removed to continue with the etching of the ITO films with a (1:1) solution of hydrochloric acid (HCL). This last process was done at 40 °C temperature for 2 min. Finally, we remove the photoresistant layer. Besides, an SMA connector was employed to connect the antenna; this connector was soldered to the transparent antenna using indium (In) solder spheres. The fabricated antenna placed onto the BUAP label is shown in Figure 4, where the transmittance of the ITO can be observed. The vector network analyzer (VectorStar, Anritsu) was used to measure the S11 from 0.5 GHz to 12 GHz. The simulated and measured return loss (S11) are shown and compared in Figure 5. The measured S11 at 5.9 GHz is −16.7 dB, which shows a good agreement with the simulated results.

Figure 4. Manufactured antenna photography.

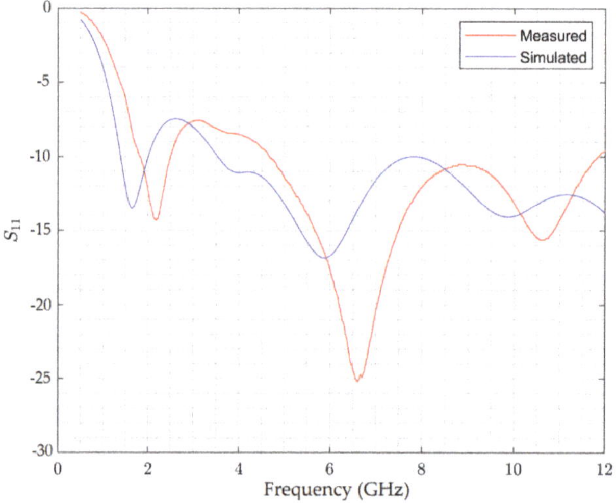

Figure 5. Comparison of simulated and measured return loss parameters.

Although a slightly detuned response is obtained from the measurement, related to fabrication flaws, each fractional bandwidth (FWB) remains, as can be seen in Table 3, showing the FBW for every band in simulation and measurement. The measured FBWs for S11 below −10 dB at the lower band is 31.34% (1.82 to 2.5 GHz), and at the upper band is 87.03% (4.66 to 11.84 GHz), which is large enough to cover the 5.9 GHz required frequency. Thus, experimental results fulfill the communication IEEE 802.11p standard for the required band.

Table 3. Simulated (S) and Measured (M) Fractional Bandwidth Comparison. FBW: Fractional Bandwidth.

	f_{range}	FBW
S	1.39–1.99	36.36%
M	1.82–2.5	31.34%
S	3.54–7.75	71.35%
M	4.66–11.84	87.03%

Figure 6 shows the simulated and measured 2-D radiation patterns of the proposed antenna at 2.1 and 5.9 GHz. At both frequencies, the H-plane radiation pattern presents omnidirectional behavior, whereas the E-plane radiation pattern is bidirectional. Moreover, to complete the characterization of the antenna, the peak gain of the antenna was simulated and measured at 2.1 and 5.9 GHz—Table 4 summarizes the obtained peak gains. The measured radiation patterns and peak gains are in concordance with the simulate results. Besides, the antenna efficiency at 2.1 and 5.9 GHz was simulated; the obtained values are 10% and 15%, respectively. These gain peaks and efficiency values are in agreement with those reported for other antennas that use ITO as a conductive film [43].

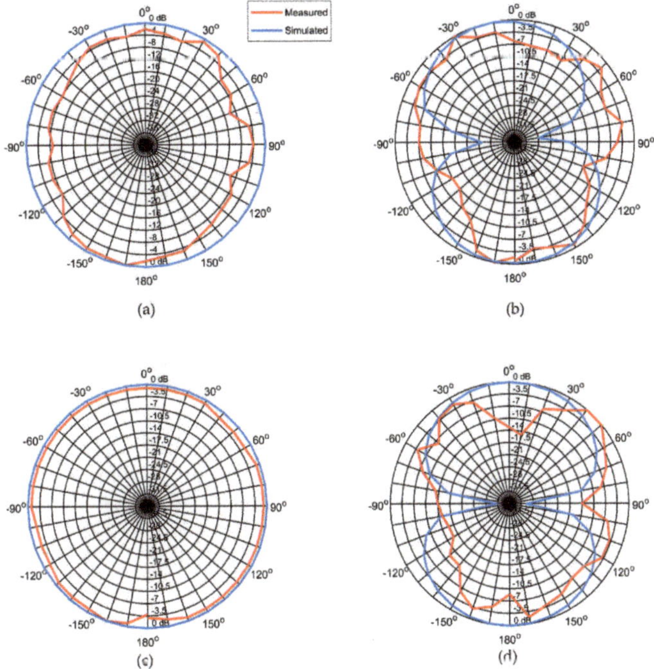

Figure 6. Simulated and measured radiation patterns: (**a**) H-plane patterns at 5.9 GHz; (**b**) E-plane patterns at 5.9 GHz; (**c**) H-plane patterns at 2.1 GHz; (**d**) E-plane patterns at 2.1 GHz.

Table 4. Simulated (S) and Measured (M) Peak Gain.

	2.1 GHz	5.9 GHz
S	−0.167 dB	−2.12 dB
M	−0.145 dB	−3.9 dB

Furthermore, the current distribution for the transparent antenna was obtained by simulation at 2.1 and 5.9 GHz, and it is presented in Figure 7. From Figure 7a, it is observed that the distribution current at 2.1 GHz is more intense on the feedline and, at the bottom of the middle path. Another significant amount of current is around the geometry of the antenna. On the other hand, at 5.9 GHz, the current distribution is mainly on the feedline, and at the middle patch. However, there is a significant amount of current between the resonators and at the largest resonator edges, as shown in Figure 7b.

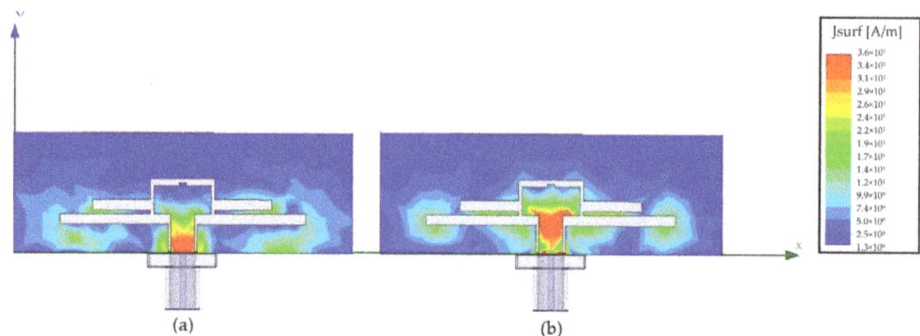

Figure 7. Simulated distribution current at (**a**) 2.1 GHz and (**b**) 5.9 GHz.

Finally, Table 5 presents a comparison of the proposed antenna with other transparent antennas. From the table, it can be noted that the antennas manufactured with ITO [22–24] do not report a fair agreement between the simulated and measured values of the reflection coefficient throughout the entire frequency range of the study, whereas in this work we can see a reasonable concordance between simulated and measured values. The frequency shift may be due to the manufacturing process and the variations in the permittivity of the substrate.

Table 5. Comparative Analysis with Other Transparent Antennas.

Reference.	Frequencies (GHz)	S11 Min Level (dB)	Substrate//Conductive Film	Dimensions (mm)	Match between M and S of S11 *
[25]	2.19–2.58	−13.97	Sapphire//GZO	$60 \times 10 \times 0.375$	No
[29]	2–6	−25	Glass//ITO	$25 \times 25 \times 1.1$	No
[30]	1–7	<−15	Glass//ITO	$50 \times 50 \times 1.1$	No reported
[32]	4.5–7	−25	Acetate//ITO	$50 \times 51 \times 0.28$	No
[33]	2.4–2.65	−15	C58//IZTO/Ag/IZTO	23.4×2	Yes
[38]	3.89–5.97	−24	PET//AgHT-8	$58 \times 78 \times 0.9$	Yes
This work	1.8–2.5/4.66–11.84	−25	Glass//ITO	$50 \times 17 \times 1.1$	Yes

* (M) Measured, (S) Simulated.

4. Conclusions

A CPW-fed fully transparent antenna for vehicular communication standards was designed, manufactured, and measured. The crossmatch technique was used for the impedance matching to achieve the target frequencies through parametric simulation. The antenna was designed and fabricated using commercial ITO-coated glass with a sheet resistance of 8 ohms/square. Opposite to other works that also use conductive films with a high sheet resistance, the experimental parameters obtained showed a behavior following the one simulated. Despite the frequency shift in the measured reflection coefficient, the fabricated antenna covers the range of 1.82–2.5 GHz at the lower band and 4.66–11.84 GHz at the upper band for 2.1 GHz and 5.9 GHz requirements, respectively. Those ranges comply with the IEEE 802.11p standard and LTE sub 6 GHz communication bands for vehicular communications applications. The proposed antenna achieves an omnidirectional radiation H-plane patterns. As expected, the peaks gain of the antenna, in the desired frequencies, were negative. Additionally, the proposed antenna presents low radiation efficiency under 15%. These behaviors are due to the sheet resistance of the ITO film, and it can be improved by increasing the thickness of the ITO film, but it would reduce the transparency of the antenna.

Author Contributions: Conceptualization, J.I.T.-F., R.T.-M., J.M.M.-P., M.A.V.-A., E.I.T.-F., M.L.-L., and E.C.-B.; methodology, J.I.T.-F., R.T.-M., and E.C.-B.; software, J.I.T.-F., R.T.-M., M.L.-L., and E.C.-B.; validation, J.I.T.-F., R.T.-M., and E.C.-B.; formal analysis, J.I.T.-F., R.T.-M., J.M.M.-P., E.I.T.-F., M.L.-L., and E.C.-B.; investigation, J.I.T.-F.,

R.T.-M., and E.C.-B.; measurements, J.I.T.-F., R.T.-M., and E.C.-B.; data curation, J.I.T.-F., R.T.-M., and E.C.-B.; writing—original draft preparation, J.I.T.-F., R.T.-M., and E.C.-B.; writing—review and editing, J.I.T.-F., R.T.-M., J.M.M.-P., E.I.T.-F., M.L.-L., and E.C.-B.; supervision, R.T.-M. and J.M.M.-P.; project administration, J.I.T.-F., and R.T.-M.; All authors have read and agreed to the published version of the manuscript.

Funding: This research received no external funding.

Acknowledgments: With thanks to the laboratory of characterization of systems based on microwaves at FCE-BUAP where all experimental characterizations were carried out. J.

Conflicts of Interest: The authors declare no conflict of interest.

References

1. Koga, Y.; Kai, M. A transparent double folded loop antenna for IoT applications. In Proceedings of the 2018 IEEE-APS Topical Conference on Antennas and Propagation in Wireless Communications (APWC), Cartagena des Indias, Colombia, 10–14 September 2018; pp. 762–765.
2. Jain, B.; Brar, G.; Malhotra, J.; Rani, S. A novel approach for smart cities in convergence to wireless sensor networks. *Sustain. Cities Soc.* **2017**, *35*, 440–448. [CrossRef]
3. Kirimtat, A.; Krejcar, O.; Kertesz, A.; Tasgetiren, M.F. Future trends and current state of smart city concepts: A survey. *IEEE Access* **2020**, *8*, 86448–86467. [CrossRef]
4. Alavi, A.H.; Jiao, P.; Buttlar, W.G.; Lajnef, N. Internet of things-enabled smart cities: State-of-the-art and future trends. *Measurement* **2018**, *129*, 589–606. [CrossRef]
5. Bibri, S.E.; Krogstie, J. Smart sustainable cities of the future: An extensive interdisciplinary literature review. *Sustain. Cities Soc.* **2017**, *31*, 183–212. [CrossRef]
6. Wang, J.; Jiang, C.; Zhang, K.; Quek, T.Q.S.; Ren, Y.; Hanzo, L. Vehicular sensing networks in a smart city: Principles, technologies and applications. *IEEE Wirel. Commun.* **2018**, *25*, 122–132. [CrossRef]
7. Matyakubov, M.; Rustamova, O. Development of smart city model: Smart bus system. In Proceedings of the 2019 International Conference on Information Science and Communications Technologies (ICISCT), Tashkent, Uzbekistan, 9–10 March 2019; pp. 1–5.
8. Yang, L.; Mo, T.; Li, H. Research on V2V communication based on peer to peer network. In Proceedings of the 2018 International Conference on Intelligent Autonomous Systems (ICoIAS), Singapore, 1–3 March 2018; pp. 105–110.
9. Alami, A.J.; El-Sayed, K.; Al-Horr, A.; Artail, H.; Guo, J. Improving the Car GPS accuracy using V2V and V2I communications. In Proceedings of the 2018 IEEE International Multidisciplinary Conference on Engineering Technology (IMCET), Beirut, Lebanon, 14–16 November 2018; pp. 1–6.
10. Franzo, S.; Latilla, V.M.; Longo, M.; Bracco, S. Towards the new concept of smart roads: Regulatory framework and emerging projects overview. In Proceedings of the 2018 International Conference of Electrical and Electronic Technologies for Automotive, Milan, Italy, 9–11 July 2018; pp. 1–6.
11. Jiang, D.; Delgrossi, L. Towards an international standard for wireless access in vehicular environments. In Proceedings of the VTC Spring 2008—IEEE Vehicular Technology Conference, Singapore, 11–14 May 2008; pp. 2036–2040.
12. Roque-Cilia, S.; Tamariz-Flores, E.I.; Torrealba-Meléndez, R.; Covarrubias-Rosales, D.H. Transport tracking through communication in WDSN for smart cities. *Measurement* **2019**, *139*, 205–212. [CrossRef]
13. *IEEE Standard for Information Technology—Local and Metropolitan Area Networks—Specific Requirements—Part 11: Wireless LAN Medium Access Control (MAC) and Physical Layer (PHY) Specifications Amendment 6: Wireless Access in Vehicular Environments*; IEEE: New York, NY, USA, 2010.
14. Ucar, S.; Ergen, S.C.; Ozkasap, O. Multihop-cluster-based IEEE 802.11p and LTE hybrid architecture for VANET safety message dissemination. *IEEE Trans. Veh. Technol.* **2016**, *65*, 2621–2636. [CrossRef]
15. Arena, F.; Pau, G.; Severino, A. A Review on IEEE 802.11p for intelligent transportation systems. *J. Sens. Actuator Netw.* **2020**, *9*, 22. [CrossRef]
16. Choi, J.-Y.; Jo, H.-S.; Mun, C.; Yook, J.-G. Preamble-based adaptive channel estimation for IEEE 802.11p. *Sensors* **2019**, *19*, 2971. [CrossRef]
17. Klapez, M.; Grazia, C.A.; Casoni, M. Minimization of IEEE 802.11p packet collision interference through transmission time shifting. *J. Sens. Actuator Netw.* **2020**, *9*, 17. [CrossRef]

18. Xie, Y.; Ho, I.W.; Magsino, E.R. The modeling and cross-layer optimization of 802.11p VANET Unicast. *IEEE Access* **2018**, *6*, 171–186. [CrossRef]

19. Syfullah, M.; Lim, J.M.-Y. Data broadcasting on Cloud-VANET for IEEE 802.11p and LTE hybrid VANET architectures. In Proceedings of the 3rd International Conference on Computational Intelligence & Communication Technology (CICT), Ghaziabad, India, 10–13 February 2017; pp. 1–6.

20. Arora, A.; Mehra, A.; Mishra, K.K. Vehicle to vehicle (V2V) VANET based analysis on waiting time and performance in LTE network. In Proceedings of the 3rd International Conference on Trends in Electronics and Informatics (ICOEI), Tirunelveli, India, 23–25 April 2019; pp. 482–489.

21. *IEEE Standard for Definitions of Terms for Antennas*; IEEE: New York, NY, USA, 1973.

22. Duraj, D.; Rzymowski, M.; Nyka, K.; Kulas, L. ESPAR Antenna for V2X applications in 802.11p frequency band. In Proceedings of the 13th European Conference on Antennas and Propagation (EuCAP), Krakow, Poland, 31 March–5 April 2019; pp. 1–4.

23. Condo Neira, E.; Carlsson, J.; Karlsson, K.; Ström, E.G. Combined LTE and IEEE 802.11p antenna for vehicular applications. In Proceedings of the 9th European Conference on Antennas and Propagation (EuCAP), Lisbon, Portugal, 12–17 April 2015; pp. 1–5.

24. Madhav, B.; Anilkumar, T.; Kotamraju, K. Transparent and conformal wheel-shaped fractal antenna for vehicular communication applications. *AEU Int. J. Electron. Commun.* **2018**, *91*, 1–10. [CrossRef]

25. Green, R.B.; Toporkov, M.; Ullah, M.D.B.; Avrutin, V.; Ozgur, U.; Morkoc, H.; Topsakal, E. An alternative material for transparent antennas for commercial and medical applications. *Microw. Opt. Technol. Lett.* **2017**, *59*, 773–777. [CrossRef]

26. Lee, S.; Choo, M.; Jung, S.; Hong, W. Optically transparent nano-patterned antennas: A review and future directions. *Appl. Sci.* **2018**, *8*, 901. [CrossRef]

27. Eltresy, N.A.; Elsheakh, D.N.; Abdallah, E.A.; Elhennawy, H.M. RF energy harvesting using transparent antenna for IoT application. In Proceedings of the 2019 International Conference on Innovative Trends in Computer Engineering (ITCE), Aswan, Egypt, 2–4 February 2019; pp. 287–291.

28. Cai, L. An on-glass optically transparent monopole antenna with ultrawide bandwidth for solar energy harvesting. *Electronics* **2019**, *8*, 916. [CrossRef]

29. Hakimi, S.; Rahim, S.K.A.; Abedian, M.; Noghabaei, S.M.; Khalily, M. CPW-fed transparent antenna for extended ultrawideband applications. *IEEE Antennas Wirel. Propag. Lett.* **2014**, *13*, 1251–1254. [CrossRef]

30. Júnior, P.F.D.S.; Freire, R.C.S.; Serres, A.; Catunda, S.Y.; Silva, P.H.D.F. Bioinspired transparent antenna for WLAN application in 5 GHz. *Microw. Opt. Technol. Lett.* **2017**, *59*, 2879–2884. [CrossRef]

31. Hong, S.; Kang, S.H.; Kim, Y.; Jung, C.W. Transparent and flexible antenna for wearable glasses applications. *IEEE Trans. Antennas Propag.* **2016**, *64*, 2797–2804. [CrossRef]

32. Hautcoeur, J.; Colombel, F.; Castel, X.; Himdi, M.; Cruz, E.M. Radiofrequency performances of transparent ultra-wideband antennas. *Prog. Electromagn. Res. C* **2011**, *22*, 259–271. [CrossRef]

33. Martin, A.; Castel, X.; Lafond, O.; Himdi, M. Optically transparent frequency-agile antenna for X-band applications. *Electron. Lett.* **2015**, *51*, 1231–1233. [CrossRef]

34. Martin, A.; Castel, X.; Himdi, M.; Lafond, O. Mesh parameters influence on transparent and active antennas performance at microwaves. *AIP Adv.* **2017**, *7*, 085120. [CrossRef]

35. Martin, A.; Lafond, O.; Himdi, M.; Castel, X. Improvement of 60 GHz transparent patch antenna array performance through specific double-sided micrometric mesh metal technology. *IEEE Access* **2019**, *7*, 2256–2262. [CrossRef]

36. Desai, A.; Upadhyaya, T.; Patel, J.; Patel, R.; Palandoken, M. Flexible CPW fed transparent antenna for WLAN and sub-6 GHz 5G applications. *Microw. Opt. Technol. Lett.* **2020**, *62*, 2090–2103. [CrossRef]

37. Desai, A.; Upadhyaya, T.; Palandoken, M.; Patel, J.; Patel, R. Transparent conductive oxide-based multiband CPW fed antenna. *Wirel. Pers. Commun.* **2020**, *113*, 961–975. [CrossRef]

38. Kashanianfard, M.; Sarabandi, K. Vehicular optically transparent UHF antenna for terrestrial communication. *IEEE Trans. Antennas Propag.* **2017**, *65*, 3942–3949. [CrossRef]

39. Lee, J.; Lee, H.; Kim, D.; Jung, W. Transparent dual-band monopole antenna using a μ-metal mesh on the rear glass of an automobile for frequency modulation/digital media broadcasting service receiving. *Microw. Opt. Technol. Lett.* **2019**, *61*, 503–508. [CrossRef]

Appl. Sci. **2020**, *10*, 6001

40. Liu, T.-W.; Tu, W.-H. CPW-fed tri-band slot antenna with impedance matching stub. In Proceedings of the 2016 IEEE/ACES International Conference on Wireless Information Technology and Systems (ICWITS) and Applied Computational Electromagnetics (ACES), Honolulu, HI, USA, 13–17 March 2016; pp. 1–2.

41. Chen, S.-Y.; Chen, Y.-C.; Hsu, P. CPW-fed aperture-coupled slot dipole antenna for tri-band operation. *IEEE Antennas Wirel. Propag. Lett.* **2008**, *7*, 535–537. [CrossRef]

42. Smits, F.M. Measurement of sheet resistivities with the four-point probe. *Bell Syst. Tech. J.* **1958**, *37*, 711–718. [CrossRef]

43. Haraty, M.R.; Naser-Moghadasi, M.; Lotfi-Neyestanak, A.A.; Nikfarjam, A. Improving the efficiency of transparent antenna using gold nano Layer deposition. *IEEE Antennas Wirel. Propag. Lett.* **2015**, 4–7. [CrossRef]

Article

High Selectivity Slot-Coupled Bandpass Filter Using Discriminating Coupling and Source-Load Coupling

Jie Cui, Haojie Chang and Renli Zhang *

School of Electronic and Optical Engineering, Nanjing University of Science and Technology,
Nanjing 210094, China; cuijie@njust.edu.cn (J.C.); njusteo_changhj@njust.edu.cn (H.C.)
* Correspondence: zhangrenli@njust.edu.cn

Received: 15 August 2020; Accepted: 17 September 2020; Published: 28 September 2020

Abstract: A multilayer bandpass filter with high selectivity is proposed in this letter. Discriminating coupling formed by slot-coupled quarter-wavelength and half-wavelength resonators introduces a zero at $3f_0$ (f_0 is the center frequency) and the second harmonic is also suppressed due to the quarter-wavelength resonators. Owing to multilayer structure, source-load coupling is introduced to improve selectivity. Then an extra coupled line path is added with the same amplitude as the discriminating coupling path while they are out of phase. Thus signal cancellation produces three extra transmission zeros, with the selectivity and suppression performance further improved. To validate the design, a prototype bandpass filter centered at 2.49 GHz with 3 dB fractional bandwidth of 8.1% is fabricated. Both simulated and measured results are in good agreement and show good performance of the proposed bandpass filter.

Keywords: bandpass filter; discriminating coupling; high selectivity; source-load coupling

1. Introduction

Bandpass filter (BPF) is an essential component in transmitters and receivers. For anti-interference, high performance BPF with sharp selectivity as well as wide stopband is important and necessary. Due to limited and crowded spectrum nowadays, there are more and more challenges in designing such bandpass filters in modern wireless systems.

Traditionally, the selectivity of BPF can be improved by increasing the order at the expense of larger insertion loss. While quasi-elliptic function response can realize the same selectivity with fewer orders and thus quasi-elliptic filter has lower insertion loss with high selectivity [1]. In [2], stub-loaded half-wavelength resonators are electromagnetically coupled to form quasi-elliptic response. In addition to specific response functions, introducing multiple transmission zeros (TZs) near the passband can effectively improve the selectivity [3]. Source-load coupling [4,5] and mixed coupling [6] are both proved to be useful by introducing TZs at sidebands. In [7], capacitive dominant mixed coupling is used to create a lower stopband TZ while the upper stopband selectivity is enhanced by employing parallel source-load coupling. Mixed coupling is also common when designing high selectivity SIW BPFs [8–10]. Furthermore, merely several coupled line networks can introduce both transmission zeros and poles. In [11], a high selectivity bandpass filter with 5 zeros and 6 poles is obtained by six pairs of coupled lines. Another way to realize prescribed TZs is synthesis of the coupling matrix to generate the transfer and reflection polynomials for specific class of filter[12,13], and to simplify the computational complexity and eliminate redundancy, synthesis algorithms are invented to extract coupling matrix from zeros and poles[14,15].

In this paper, a multilayer bandpass filter with high selectivity is proposed. The second and third harmonics are eliminated by the quarter-wavelength resonators and discriminating coupling formed by quarter- and half-wavelength resonators respectively, resulting in a wide stopband

Appl. Sci. **2020**, *10*, 6807; doi:10.3390/app10196807

as well. This discriminating coupling path has same amplitude as an extra coupled line path between feeding ports but they are out of phase. Thus transmission zeros are produced to not only improve the selectivity but also broaden the stopband performance of the proposed BPF. Furthermore, due to multilayer structure, source-load coupling is easily achieved to improve selectivity as well. Theoretical analysis of the proposed BPF is demonstrated in detail and a prototype is fabricated to validate the design. Both simulated and measured results are in good agreement, showing the good performance of the proposed BPF.

2. Structure and Design of The BPF

The proposed BPF has two independent paths as is shown in Figure 1. Upper path I is composed of a pair of shorted quarter-wavelength resonators in the top layer and one half-wavelength resonator in the bottom layer, with two rectangular slots etched in the middle ground layer to realize slot-coupling. And this path is coupled to two quarter-wavelength short microstrip lines as exciting structure. Lower path II consists of a stub-loaded anti-parallel coupled line, connecting to the feeding ports in parallel with upper path I. Both of these two paths will be analyzed in detail next.

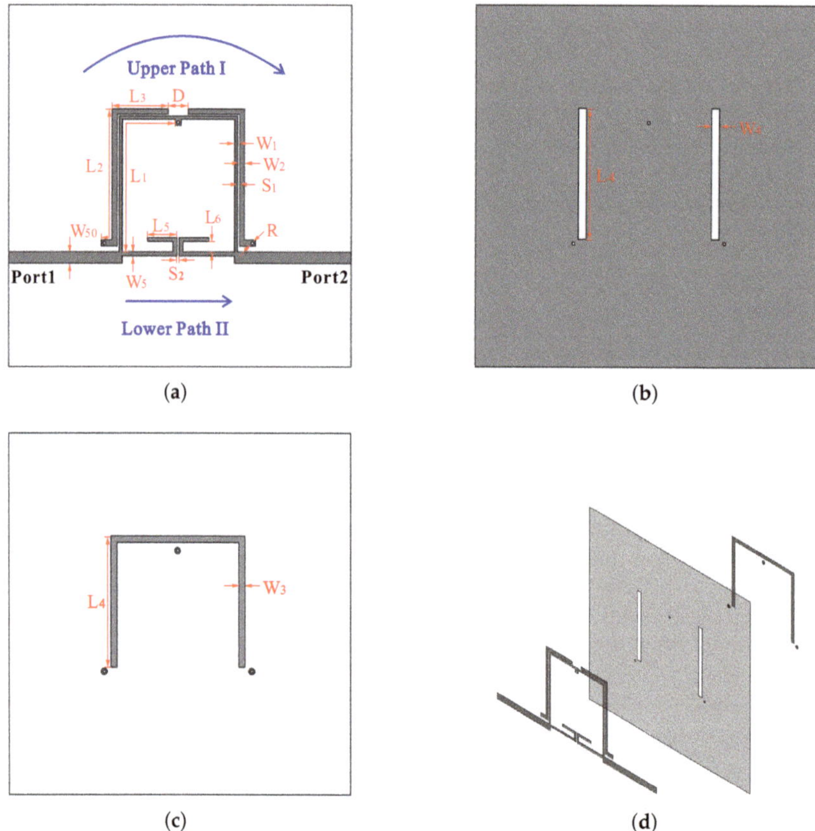

Figure 1. *Cont.*

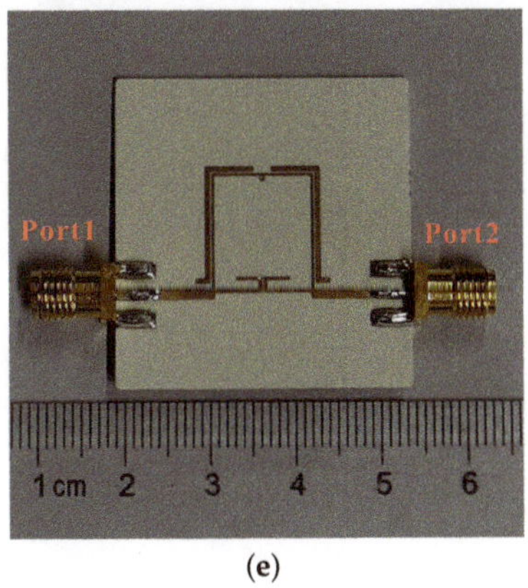

(e)

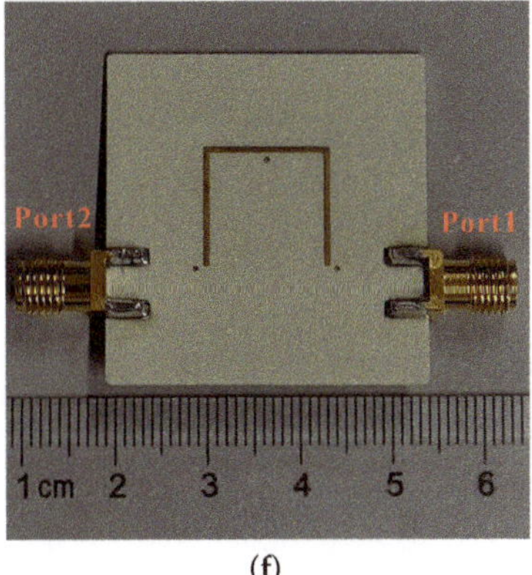

(f)

Figure 1. Configuration of the proposed bandpass filter (BPF) (**a**) Top layer. (**b**) Middle layer. (**c**) Bottom layer. (**d**) 3D view. (**e**) Front view of the proposed BPF. (**f**) Bottom view of the proposed BPF.

2.1. Analysis of Upper Path I

Figure 2 shows the equivalent circuit of upper path I. The coupling networks consisting of quarter- and half-wavelength resonators are marked as red (T1–T5), (T1+T2 and T4+T5 are quarter-wavelength, T2+T3+T4 is half-wavelength), which are excited by two shorted quarter-wavelength resonators marked as blue (T6) from port 1 and 2, respectively. T2 and T4 are two end-to-end coupled lines with even and odd impedance of Z_{e1} and Z_{o1}, electric length of θ_2. Three Microstrip lines T1, T3 and T5 are connected to coupling networks T2 and T4. T1 and T5 are connected to the shorted lines in T2 and T4 with impedance of Z_1 and electric length of θ_1 while T3 is used to connect T2 and T4 with impedance of Z_2 but two times longer than T1 or T5. The total length of T1 and T2 or T4 and T5 is quarter-wavelength ($\theta_1 + \theta_2 = \pi/2$).

To verify the analysis, we use transfer matrix ABCD which is defined as Equation (1) shows [16]:

$$\begin{pmatrix} V_1 \\ I_1 \end{pmatrix} = \begin{pmatrix} A & B \\ C & D \end{pmatrix} \begin{pmatrix} V_2 \\ -I_2 \end{pmatrix} \tag{1}$$

For transmission lines with character impedance Z_0 and electrical length θ [16]:

$$\begin{pmatrix} A & B \\ C & D \end{pmatrix} = \begin{pmatrix} cos\theta & jZ_0\sin\theta \\ jY_0\sin\theta & \cos\theta \end{pmatrix} \tag{2}$$

Thus the response of the coupling network can be calculated by multiplying the ABCD matrixes as Equation (3).

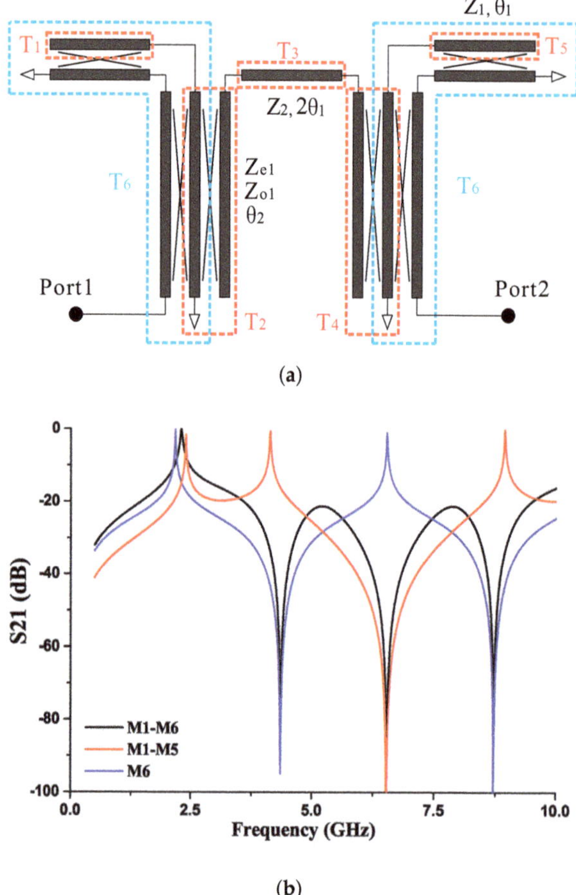

(**a**)

(**b**)

Figure 2. Equivalent circuit and MATLAB simulated results of resonators in upper path I. (**a**) Equivalent circuit of different sections in upper path I. (**b**) MATLAB simulated results of different sections in upper path I.

$$T = T_1 \times T_2 \times T_3 \times T_4 \times T_5 \tag{3}$$

where

$$T_1 = \begin{pmatrix} cos\theta_1 & jZ_1 \sin \theta_1 \\ jY_1 \sin \theta_1 & \cos \theta_1 \end{pmatrix} = T_5 \tag{4a}$$

$$T_3 = \begin{pmatrix} \cos 2\theta_1 & jZ_2 \sin 2\theta_1 \\ jY_2 \sin 2\theta_1 & \cos 2\theta_1 \end{pmatrix} \tag{4b}$$

$$T_2 = \frac{1}{ab - cf} \begin{pmatrix} a^2 - df & ac - db \\ ea - bf & ec - b^2 \end{pmatrix} \tag{4c}$$

$$T_4 = \frac{1}{ab - de} \begin{pmatrix} b^2 - ce & db - ac \\ fb - ae & fd - a^2 \end{pmatrix} \tag{4d}$$

Here a to f in Equation (4) represent formulas related to the dimensions of coupling network T2 and T4 [17] as Equation (5).

$$a = \frac{1}{2}(\cos\theta_{2e} + \cos\theta_{2o}) \tag{5a}$$

$$b = \frac{1}{2}(\cos\theta_{2e} - \cos\theta_{2o}) \tag{5b}$$

$$c = j\frac{1}{2}(Z_{e1}\cos\theta_{2e} + Z_{o1}\cos\theta_{2o}) \tag{5c}$$

$$d = j\frac{1}{2}(Z_{e1}\cos\theta_{2e} - Z_{o1}\cos\theta_{2o}) \tag{5d}$$

$$e = j\frac{1}{2}(Y_{e1}\cos\theta_{2e} + Y_{o1}\cos\theta_{2o}) \tag{5e}$$

$$f = j\frac{1}{2}(Y_{e1}\cos\theta_{2e} - Y_{o1}\cos\theta_{2o}) \tag{5f}$$

θ_{2e} and θ_{2o} are the even and odd electric lengths of slot-coupled resonators while Z_{e1}, Z_{o1} and Y_{e1}, Y_{o1} are the impedances and admittances. All of these parameters can be obtained using quasi-static analysis [18].

It is noteworthy that if the length of the slots θ_2 is two-third of quarter-wavelength, discriminating coupling is formed. According to [19], a zero at $3f_0$ will appear when discriminating coupling occurs, because the voltage distribution is odd on one line and even on the other, leading to null coupling coefficient. Figure 2 shows the simulated result using MATLAB to validate the performance of discriminating. Blue line in Figure 2 demonstrates the simulated result of coupling networks including the connecting microstrip lines (T1-T5) and it is seen that there are two modes and one zero. Thus this line validates that discriminating coupling is effective to introduce a transmission zero at $3f_0$ in this design. However, the second mode is located at around $2f_0$ because coupling network T2,T4 and microstrip line section M3 realize a half-wavelength resonator in the bottom layer as is clearly shown in Figure 1. To suppress the harmonic caused by half-wavelength resonator, quarter-wavelength shorted coupled lines (M6) is used as feeding lines to introduce another zero at $2f_0$. The red line in Figure 2 proves the function of M6.

Finally the black line in Figure 2 gives the MATLAB simulated result of the whole structure including T1-T6. It shows that the second and third harmonics are suppressed by zeros introduced by discriminating coupling and quarter-wavelength resonators, validating the theory and design very well. Also due to multilayer structure, it is easy to introduce source-load coupling using the short pins of M6 as is shown in the top layer of Figure 1. Thus a zero located at lower side-band appears and improves the selectivity. HFSS simulated results will validate the design in the next subsection.

2.2. Analysis of Lower Path Ii

The lower path II consists of a stub-loaded anti-parallel coupling section connected by microstrip lines, as is shown in Figure 3. Here the impedance and electric length of connecting microstrip lines are Z_3 and θ_3. The anti-parallel coupled line has even and odd impedance of Z_{e2} and Z_{o2} and the electric length of θ_3 with loaded stubs of Z_4 and θ_4. Formula Equation (6) gives the ABCD matrix of lower path II.

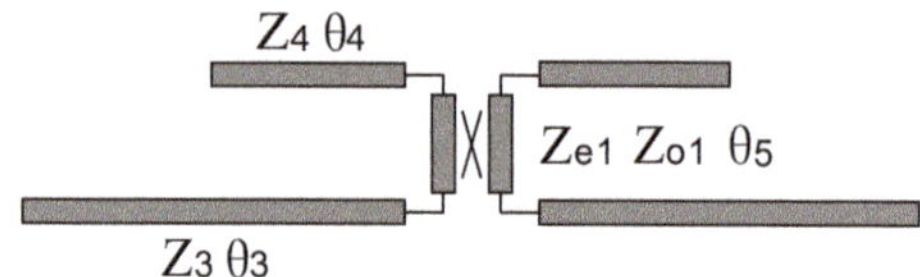

Figure 3. Equivalent circuit of anti-parallel coupling network in lower path II.

$$T = \begin{pmatrix} \cos\theta_3 & jZ_3\sin\theta_3 \\ jY_3\sin\theta_3 & \cos\theta_1 \end{pmatrix} \begin{pmatrix} A_c & B_c \\ C_c & D_c \end{pmatrix} \begin{pmatrix} \cos\theta_3 & jZ_3\sin\theta_3 \\ jY_3\sin\theta_3 & \cos\theta_1 \end{pmatrix} \tag{6}$$

A_c, B_c, C_c and D_c in Equation (6) are the ABCD matrix's elements of the stub-loaded anti-parallel coupled line. The method to calculate these elements is same as the one used for coupling networks T2 or T4 [17] as Equation (7).

$$A_c = \frac{1}{G}[(c+aZ_s)(a+eZ_s) - (d+bZ_s)(b+fZ_s)] \tag{7a}$$

$$B_c = \frac{1}{G}[(c+aZ_s)(c+aZ_s) - (d+bZ_s)(d+bZ_s)] \tag{7b}$$

$$C_c = \frac{1}{G}[(a+eZ_s)(a+eZ_s) - (b+fZ_s)(b+fZ_s)] \tag{7c}$$

$$D_c = \frac{1}{G}[(a+eZ_s)(c+aZ_s) - (b+fZ_s)(d+bZ_s)] \tag{7d}$$

where

$$G = (d+bZ_s)(a+eZ_s) - (b+fZ_s)(c+aZ_s) \tag{8}$$

and

$$Z_s = -jZ_4\cot\theta_4 \tag{9}$$

Similarly, a to f in Equation (7) and Equation (8) are same as Equation (5) but the even and odd parameters are replaced by θ_{5e}, θ_{5o}, Z_{e2}, Z_{o2}, Y_{e2} and Y_{o2}. Formula Equation (9) is the input impedance of the open stubs.

MATLAB can also be used to calculate the response of combined upper and lower path. However, the formulas will be very complicated and the simulated result will not be clear enough to demonstrate the function of lower path II. Thus full-wave simulation software Ansys HFSS 19 is used. Blue line in Figure 4 shows the HFSS simulated results of upper path I while the red line gives the ones of lower path II. As is shown in Figure 4, the upper path I has bandpass response with the second and third harmonics suppressed by TZ2 and TZ3, and another transmission zero TZ1 located at lower sideband is introduced by inductive source-load coupling as is mentioned in the last subsection. It can also be seen that the response of lower path II has same amplitude as upper path I in some frequencies but at the same time they are out of phase. So in these frequencies, transmission zeros will appear because of signal cancellation. In this design, three extra transmission zeros TZ4, TZ5 and TZ6 are introduced due to signal cancellation. TZ5 located at upper sideband is used to improve the selectivity while the lower and upper stopband is improved by TZ4 and TZ6. However, TZ2 and TZ6 are both at around $2f_0$. So overall, there are five TZs used to improve selectivity and out-of-band suppression performance.

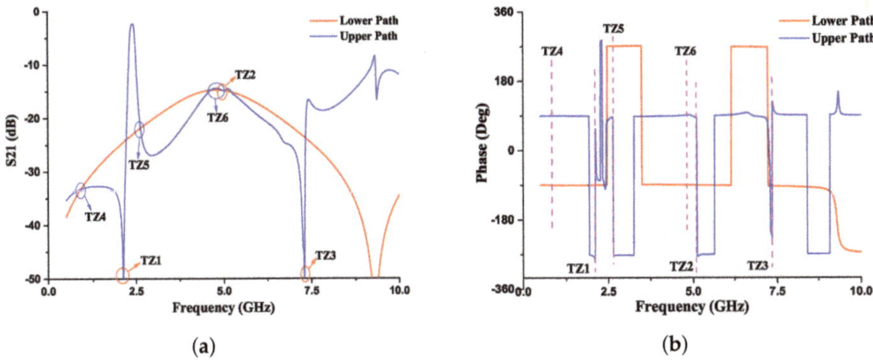

(a)   **(b)**

Figure 4. HFSS Simulated results of lower path (red line) and upper path (blue line). (**a**) Amplitude of S21 (dB). (**b**) Phase of S21 (deg)

3. Simulated and Measured Results

As shown in Figure 5, the proposed BPF is designed on two RO4350B substrate with relative permittivity of 3.66 and thickness of 0.508 mm. A 0.1 mm RO4450B prepreg is used as paste layer and so the total height of the PCB is 1.116 mm (0.508 + 0.1 + 0.508). The final parameters are: $W_1 = 0.3$, $W_2 = 0.5$, $W_3 = 0.6$, $W_4 = 0.8$, $W_5 = 0.4$, $L_1 = 18.2$, $L_2 = 13.8$, $L_3 = 5.7$, $L_4 = 12.7$, $L_5 = 3$, $L_6 = 1$, $D = 2$, $S_1 = S_2 = 0.2$, $R = 0.15$, $W_{50} = 1.1$ (unit: mm). To validate the design, a prototype is fabricated and measured using Keysigth ENA network analyser E5671C. The photograph of the proposed bandpass filter is shown in Figure 1.

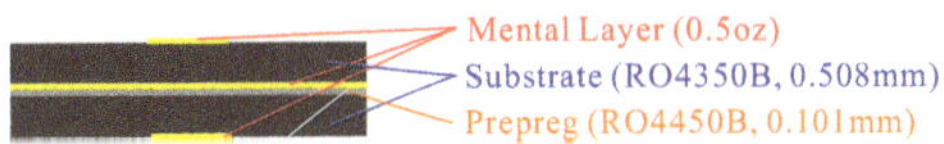

Figure 5. laminate layer definition of the fabricated PCB.

Figure 6 illustrates the simulated and measured responses. The measured 3 dB bandwidth is from 2.39 GHz to 2.59 GHz (8.1%) centered at 2.49 GHz with return loss better than -12 dB within the passband. Due to mechanical fabrication error and permittivity difference between simulation and production, the center frequency is a little higher for the measurement results. Also the insertion loss is 3.1 dB, larger than simulated one due to 0.6 dB SMA connector loss. The positions of 5 TZs are in agreement with the simulated ones with a little shift. The depicted discrepancies could be from the fabrication tolerance in the etching process. Due to these 5 TZs, the lower sideband selectivity of the proposed BPF is calculated by 3 dB and 20 dB amplitude response with respect to its frequency point:

$$Selectivity_{lower\ sideband} = \left|\frac{3-20}{f_3 - f_{20}}\right| = \frac{17}{2.39 - 2.3} = 188.8\,\text{dB/GHz} \tag{10}$$

and the upper sideband selectivity:

$$Selectivity_{upper\ sideband} = \left|\frac{3-20}{f_3 - f_{20}}\right| = \frac{17}{2.65 - 2.59} = 288.3\,\text{dB/GHz} \tag{11}$$

so that the shape factor is $BW_{20}/BW_3 = 1.75$ (BW represents bandwidth). And -15 dB suppression is from 2.64 to 8.84 GHz. The overall size is $0.2\lambda_g \times 0.22\lambda_g$ (λ_g is the wavelength at center frequency). Table 1 tabulates the performance comparisons with some previous works. Here CF represents the center frequency of the filter, and FBW is the fractional bandwidth which is calculated by passband (3 dB bandwidth) divide center frequency BW_{3dB}/f_0, and N is the number of TZs, and metal layers are the number of metal used to form the filter. TZ@sideband is the frequency of sideband transmission zeros divide center frequency $f_{sideband\ TZ} / f_0$.

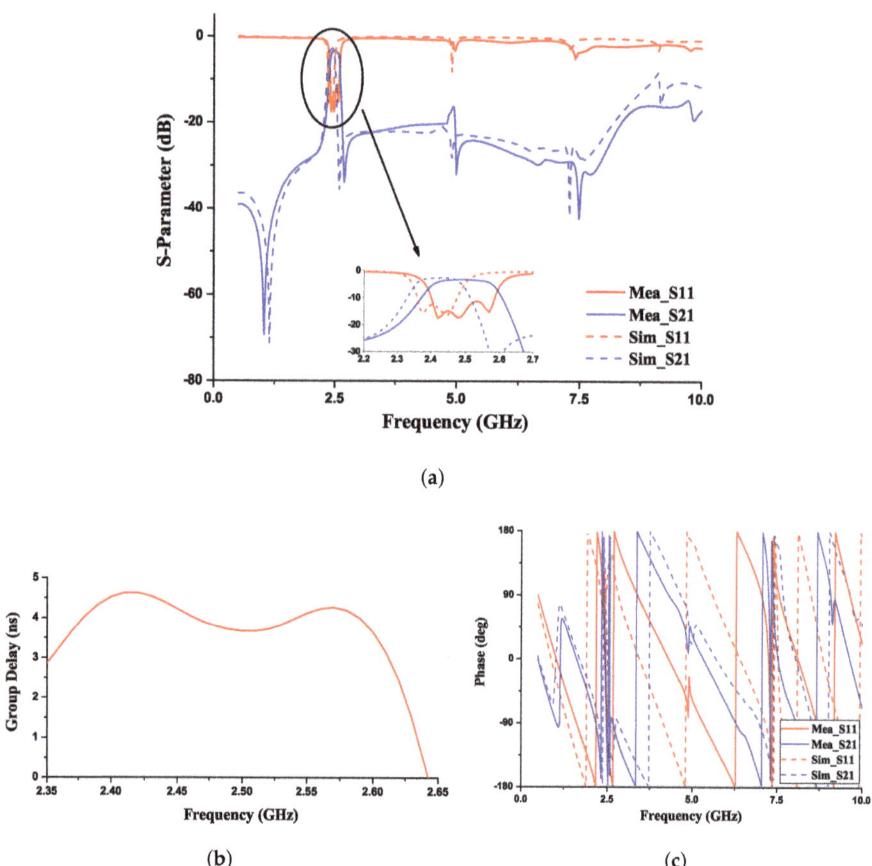

(a)

(b)   (c)

Figure 6. Simulated and measured results of the proposed BPF (solid lines: measured results; dash lines: simulated results) (**a**) Simulated and measured amplitude of S parameters (dB). (**b**) Measured group delay (ns). (**c**) Simulated and measured phase of S parameters (deg).

Table 1. Performance Comparisons with previous works

Ref	CF (GHz)	FBW (%)	N	TZ@Sizeband	Size ($\lambda_g * \lambda_g$)	Mental Layers
[1]	2	10	2	0.85 , 1.15	0.79 × 0.06	1
[4]	2.4	3.5	3	0.83 , 1.08	-	1
[7]	2.75	3.6	4	0.89 , 1.16	0.18×0.29	1
[9]	8.7	16	3	0.99 , 1.16	0.29 × 0.57	1
[10]	13.53	3.9	3	0.97 , 1.04	0.84 × 0.84	1
This work	**2.48**	**8.1**	**5**	**0.84 , 1.08**	**0.20 × 0.22**	**3**

4. Conclusions

A multilayer bandpass filter with high selectivity and wide stopband is proposed in this letter. TZs introduced by anti-parallel coupling network as well as discriminating coupling between quarter- and half-wavelength resonators improve the selectivity and out-of-band suppression of the BPF.

To verify the design, a prototype BPF is fabricated centered at 2.49 GHz. Both simulated and measured results manifest the performance with good selectivity and extended stopband. With these features, The proposed BPF is attractive in modern wireless system.

Author Contributions: Conceptualization, J.C. and H.C.; methodology, J.C.; software, H.C.; validation, J.C., H.C.and R.Z.; formal analysis, J.C.; investigation, H.C.; resources, R.Z.; data curation, J.C.; Writing—Original draft preparation, H.C.; Writing—Review and editing, R.Z.; funding acquisition, J.C. All authors have read and agreed to the published version of the manuscript.

Funding: This work has been supported by National Natural Science Foundation of China 62001232, 61971224 and Jiangsu Provincial Natural Science Foundation under Grants Nos. BK20180457.

Conflicts of Interest: The authors declare no conflict of interest.

References

1. Chen, C.J. A Coupled-Line Coupling Structure for the Design of Quasi-Elliptic Bandpass Filters. *IEEE Trans. Microw. Theory Tech.* **2018**, *66*, 1921–1925. [CrossRef]
2. Virdee, B.S.; Riaz, M.; Shukla, P.; Onadim, M.; Ouazzane, K. Wideband microstrip quasi-elliptic function bandpass filter with high out-of-band rejection. *Microw. Opt. Technol. Lett.* **2019**, *61*, 1993–1998. [CrossRef]
3. Kim, P.; Chaudhary, G.; Jeong, Y. Wide-stopband and high selectivity step impedance resonator bandpass filter using T-network and antiparallel coupled line. *IET Microw. Antennas Propag.* **2019**, *13*, 1916–1920. [CrossRef]
4. Deng, H.W.; Liu, F.; Xu, T.; Sun, L.; Xue, Y.F. Compact and high selectivity dual-mode microstrip BPF with frequency-dependent source-load coupling. *Electron. Lett.* **2018**, *54*, 219–221. [CrossRef]
5. Peng, B.; Li, S.; Zhu, J.; Zhang, Q.; Deng, L.; Zeng, Q.; Gao, Y. Wideband Bandpass Filter with High Selectivity Based on Dual-Mode DGS Resonator. *Microw. Opt. Technol. Lett.* **2016**, *58*, 2300–2303. [CrossRef]
6. Killamsetty, V.S.; Mukherjee, B. Compact Selective Bandpass Filter With Wide Stopband for TETRA Band Applications. *IEEE Trans. Compon. Packag. Manuf. Technol.* **2018**, *8*, 653–659. [CrossRef]
7. Anwar, M.S.; Cao, Q.; Burney, S.A. High selectivity quarter-wavelength resonator bandpass filter utilizing source-load coupling. *Microw. Opt. Technol. Lett.* **2020**, *62*, 1176–1182. [CrossRef]
8. Saghati, A.P.; Saghati, A.P.; Entesari, K. Ultra-Miniature SIW Cavity Resonators and Filters. *IEEE Trans. Microw. Theory Tech.* **2015**, *63*, 4329–4340. [CrossRef]
9. He, Z.; You, C.J.; Leng, S.; Li, X.; Huang, Y.M. Compact Bandpass Filter with High Selectivity Using Quarter-Mode Substrate Integrated Waveguide and Coplanar Waveguide. *IEEE Microw. Wirel. Compon. Lett.* **2017**, *27*, 809–811. [CrossRef]
10. Liu, Q.; Zhou, D.; Shi, J.; Hu, T. High-selective triple-mode SIW bandpass filter using higher-order resonant modes. *Electron. Lett.* **2020**, *56*, 37–39. [CrossRef]
11. Zhang, F.; Xu, K.D. High-selectivity bandpass filter using six pairs of quarter-wavelength coupled lines. *Electron. Lett.* **2019**, *55*, 544–546. [CrossRef]
12. Cameron, R.J. Advanced Coupling Matrix Synthesis Techniques for Microwave Filters. *IEEE Trans. Microw. Theory Tech.* **2003**, *51*, 1–10. [CrossRef]
13. Muller, A.A.; Moldoveanu, A.; Asavei, V.; Sanabria-Codesal, E.; Favennec, J.F. Lossy coupling matrix filter synthesis based on hyperbolic reflections. In Proceedings of the IEEE MTT-S International Microwave Symposium (IMS), San Francisco, CA, USA, 22–27 May 2016.
14. Muller, A.A.; Sanabria-Codesal, E.; Lucyszyn, S. Computational Cost Reduction for N+2 Order Coupling Matrix Synthesis Based on Desnanot-Jacobi Identity. *IEEE Access* **2016**, *4*, 10042–10050. [CrossRef]
15. Snyder, R.V; Mortazawi, A.; Hunter, I.; Bastioli, S.; Macchiarella, G.; Wu, K. Present and Future Trends in Filters and Multiplexers. *IEEE Trans. Microw. Theory Tech.* **2015**, *63*, 3324–3360. [CrossRef]
16. Pozar, D.M. THE TRANSMISSION (ABCD) MATRIX. In *Microwave Engineering*; John Wiley & Sons, Inc.: Hoboken, NJ, USA, 2011; pp. 188–194.
17. Zysman, G.I.; Johnson, A.K. Coupled transmission line networks in an inhomogeneous dielectric medium. *IEEE Trans. Microw. Theory Tech.* **1969**, *17*, 753–759. [CrossRef]

18. Wong, M.F.; Hanna, V.F.; Picon, O.; Baudrand, H. Analysis and design of slot-coupled directional couplers between double-sided substrate microstrip lines. *IEEE Trans. Microw. Theory Tech.* **1991**, *29*, 2123–2129. [CrossRef]

19. Li, Y.C.; Zhang, X.Y.; Xue, Q.; Baudrand, H. Bandpass Filter Using Discriminating Coupling for Extended Out-of-Band Suppression. *IEEE Microw. Wirel. Compon. Lett.* **2010**, *20*, 369–371. [CrossRef]

Article

Computational Characterization of Microwave Planar Cutoff Probes for Non-Invasive Electron Density Measurement in Low-Temperature Plasma: Ring- and Bar-Type Cutoff Probes

Si Jun Kim [1,2], **Jang Jae Lee** [2], **Young Seok Lee** [2], **Hee Jung Yeom** [3], **Hyo Chang Lee** [3], **Jung-Hyung Kim** [3] and **Shin Jae You** [2,4,*]

[1]   Nanotech Optoelectronics Research Center, Yongin Gyonggi Province 16882, Korea; sjk@o.cnu.ac.kr
[2]   Applied Physics Lab for PLasma Engineering (APPLE), Department of Physics, Chungnam National University, Daejeon 34134, Korea; leejj3800@o.cnu.ac.kr (J.J.L.); lerounsukre@o.cnu.ac.kr (Y.S.L.)
[3]   Korea Research Institute of Standards and Science, Daejeon 34113, Korea; yeom9744@kriss.re.kr (H.J.Y.); LHC@kriss.re.kr (H.C.L.); jhkim86@kriss.re.kr (J.-H.K.)
[4]   Institute of Quantum Systems (IQS), Chungnam National University, Daejeon 34134, Korea
*   Correspondence: sjyou@cnu.ac.kr

Received: 14 August 2020; Accepted: 28 September 2020; Published: 12 October 2020

Abstract: The microwave planar cutoff probe, recently proposed by Kim et al. is designed to measure the cutoff frequency in a transmission (S_{21}) spectrum. For real-time electron density measurement in plasma processing, three different types have been demonstrated: point-type, ring-type (RCP), and bar-type (BCP) planar cutoff probes. While Yeom et al. has shown that the RCP and BCP are more suitable than the point-type probe for process monitoring, the basic characteristics of the ring- and bar-type probes have yet to be investigated. The current work includes a computational characterization of a RCP and BCP with various geometrical parameters, as well as a plasma parameter, through a commercial three-dimensional electromagnetic simulation. The parameters of interest include antenna size, antenna distance, dielectric thickness of the transmission line, and input electron density. Simulation results showed that the RCP has several resonance frequencies originating from standing-wave resonance in the S_{21} spectrum that the BCP does not. Moreover, the S_{21} signal level increased with antenna size and dielectric thickness but decreased with antenna distance. Among the investigated parameters, antenna distance was found to be the most important parameter to improve the accuracy of both RCP and BCP.

Keywords: plasma diagnostics; electron density measurement; planar microwave cutoff probe; bar-type cutoff probe; ring-type cutoff probe; computational characterization

---

1. Introduction

Consisting of charged particles (electrons and ions) and neutral particles (atoms, molecules, radicals, excited and metastable species), plasma is controllable via electrostatic and electromagnetic fields [1]. Its application covers various fields such as material fabrication, nuclear fusion, and medical, environmental, and aerospace industries [1,2]. In particular, plasma is one of the key factors in semiconductor and display fabrication, since both its physical (energetic ions) and chemical (reactive radicals) properties can be exploited.

Advanced process control (APC) refers to the fine-tuning of plasma processing based on real-time feedback signals from various plasma process monitoring devices. In fabrication fields, recent demands for sub-nanometer patterning and high aspect ratio- and atomic scale-etching and deposition [3–5]

have greatly increased process difficulty, leading to wide interest in APC and real-time plasma process monitoring. The latter has contributed to productivity improvements through providing feedback signals to process control units; the feedback signals are produced via the gathering and post-processing of myriad monitoring parameters, such as plasma emission light, voltage and current of the electrode, antenna, electrostatic chuck, chamber pressure, gas flow rate, and plasma parameters (electron density, electron temperature, etc.) [4]. Among these monitoring parameters, electron density is one of the crucial factors because it is directly related to processing time and quality [1,2,6].

Many diagnostic techniques including electrical, laser, optical, and microwave methods have been developed to measure electron density. Examples include the Langmuir probe, laser Thomson scattering diagnostics, optical emission and absorption spectroscopy, and microwave probes. Most of these approaches, though, are not suitable for plasma process monitoring; the Langmuir probe cannot operate under conditions with probe tip contamination (especially in the deposition process), laser Thomson scattering diagnostics is highly sensitive to the environment and requires quite a large space, and the optical emission and absorption method operates only within a narrow window [7].

On the other hand, the microwave method has attracted great attention for application to plasma process monitoring since microwave probes are not affected by probe tip contamination, afford high measurement accuracy [8,9], and further, the required microwave power is small enough to not disturb the plasma. One drawback, though, is that microwave probes provide few parameters, namely electron density and temperature. Nowadays, research is focused on the development of a planar and compact microwave probe for real-time plasma process monitoring, since a small and planar probe can be embedded into a wafer chuck or chamber wall, and is therefore non-invasive. Variations such as the curling probe (CP), the planar multipole resonance probe (pMRP), and the planar cutoff probe have been developed and are currently under improvement via commercial three-dimensional (3D) electromagnetic simulation (CST Microwave Studio [10]), as well as experimental validation. Here, the computer simulation approach is quite simple and economical for the optimization and analysis of microwave probes.

Ogawa et al. developed the CP to measure the shift of standing wave (SW) resonance frequency caused by plasma in a reflection microwave frequency spectrum (S_{11}) [11]. The authors optimized the CP via computer simulation [11], as well as physical modeling [12,13]. They have recently demonstrated the in situ simultaneous measurement of the thickness of a deposition film and electron density using two CPs [14,15]. Developed by Schultz et al., the pMRP uses the resonance characteristics of the probe itself and can measure muti-resonance frequencies in S_{11} [16]. The antenna of the pMRP consists of two semi-circle planar plates with a dielectric cover. The authors also optimized the pMRP via CST Microwave Studio [16], as well as physical modeling [17], and practically proved the feasibility of the probe to be mounted on the chamber wall for industrial plasma processing.

Kim et al. developed the first planar cutoff probe, which measures the cutoff frequency in a transmission microwave frequency spectrum (S_{21}), analyzed the basic characteristics of the point-type probe, and optimized it via CST Microwave Studio [18]. They also fabricated the probe and firstly demonstrated its operation. Advanced types of the planar cutoff probe were subsequently developed by Yeom et al.: the ring-type planar cutoff probe (RCP) and bar-type planar cutoff probe (BCP) [19]. In [19], the authors simply compared the three designs in terms of the Q-factor of the cutoff peak using CST Microwave Studio and fabricated the BCP based on their simulation results. The BCP was then successfully applied to real-time plasma process monitoring. The basis of the planar cutoff probes is the cutoff probe, which, despite being an invasive type of probe, is considered as one of the most accurate diagnostic methods among microwave probes [20]. Therefore, it is believed that the three types of planar cutoff probes will be able to show high measurement accuracy compared to other planar microwave probes. Despite the promising potential and the fact that the RCP and BCP are more suitable for plasma process monitoring than the point-type probe [19], the basic characteristics of the ring- and bar-type probes have not been investigated yet. Hence, this paper characterizes the RCP and BCP considering various geometrical parameters, as well as a plasma parameter through the CST

Microwave Studio. The main parameters include antenna size, antenna distance, dielectric thickness, and input electron density.

This paper consists of three parts. Section 2 gives the simulation details, such as simulation method, geometry, boundary conditions, and materials. Section 3 presents the simulation results and analysis of each case. Finally, Section 4 summarizes the paper.

2. Simulation Details

The high-frequency time domain solver in CST Microwave Suite was adopted in this study. This solver used the finite-difference time-domain method to solve Maxwell's equations in 3D space, typically in the microwave range. Although this software did not solve the basic plasma equations self-consistently, compared to fluid plasma simulation [21] or particle-in-cell simulation [22], it was a useful tool to study the characteristics of the microwave probes mentioned in Section 1. Since electron density was a controllable input parameter in this simulation, the software allowed us to easily establish the ideal measurement accuracy by defining it as the deviation of the output electron density from the input electron density.

Besides, it should be noted that this simulation did not need to define the plasma source, such as a capacitively or inductively coupled plasma source, since plasma is regarded as a dispersive dielectric material. Such a material can be represented by the Drude model, in which the plasma dielectric constant (ϵ_p) is given by $\epsilon_p = \epsilon_0\left(1 - \left(2\pi f_{pe}\right)^2/\{\omega(\omega - jv_m)\}\right)$, where ϵ_0 is the vacuum dielectric constant, $f_{pe}\left(= 8980\,\sqrt{n_{e,input}}\right)$ is the plasma oscillation frequency, ω is microwave frequency, v_m is electron-neutral collision frequency, and $n_{e,input}$ is the input electron density. Here, the v_m term only includes the momentum transfer collision between electrons and Argon atoms at an electron temperature of 2.0 eV, with Maxwellian distribution for simplicity. The ϵ_p is related to the complex wavenumber (k) from the dispersion relation for electromagnetic waves ($k = \omega/c\,\sqrt{\epsilon_p}$) [5]. The real and imaginary parts of k are related to the refractive index and attenuation constant, respectively.

Figure 1a,b show schematic diagrams of the top and cross-sections of the RCP and BCP, respectively, which are embedded in a cylindrical holder. Both probes consisted of radiating and detecting antennae; the RCP had a point-type radiating antenna and ring-type detecting antenna, while the BCP had bar-type radiating and detecting antennae. All antennae had 1.0 mm height, were insulated by a dielectric of height 2.0 mm, and were connected with a 50 Ω coaxial line. The plasma was a cylinder of 330 mm diameter and 40 mm height, and was positioned with an interlayer (sheath) distance of 5.0 mm from the RCP and BCP, as shown in Figure 1c; here, the simulation considers the sheath as a vacuum dielectric material ($\epsilon_{sheath} = \epsilon_0$).

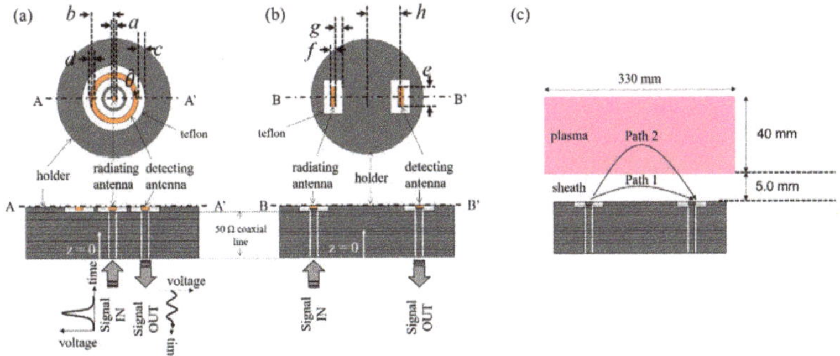

Figure 1. Schematics of the (**a**) ring-type cutoff probe (RCP) and (**b**) bar-type cutoff probe (BCP), with various geometrical parameters: *a*—antenna diameter; *b,h*—antenna distances; *c,g*—dielectric thickness; *e*—antenna length; *d,f*—antenna widths. (**c**) Geometry of the plasma and sheath.

Input and output ports were located at the ends of the 50 Ω coaxial lines. First, a Gaussian voltage pulse ($V_{in}(t)$) with 1 V maximum, including frequencies from zero to 10 GHz, was applied to the input port and radiated by the radiating antenna. Then, a fraction of the pulses ($V_{out}(t)$) entered the output port via the detecting antenna. The S_{21} is defined as $S_{21} = 20 \log_{10}(V_{in}(f)/V_{out}(f))$ (dB), where $V_{in}(f)$ and $V_{out}(f)$ are the processed data through a fast Fourier transform of $V_{in}(t)$ and $V_{out}(t)$.

All simulations were conducted in the open boundary condition except for the $z = 0$ plane where the input and output ports are located. On this plane, the ground boundary condition (zero tangential electric field) was applied.

Figure 1a,b represent the geometrical simulation variables (*a* to *h*), and Table 1 lists the simulation conditions. For simple expression, except for antenna diameter (*a*) and antenna length (*e*), all parameters were normalized as α ($\equiv d/a$ or f/e), β ($\equiv b/a$ or h/e), and γ ($\equiv c/a$ or g/e). In each simulation case, α changed from 1 to 4 for the RCP, and β and γ changed from 4.5 to 7 and 0.5 to 1.5 for both probes, respectively. Additionally, the normalized factors (*a* and *e*) changed from 2.0 to 3.0 mm. Otherwise, the plasma parameter (input electron density) also changed from 1×10^9 to 1×10^{11} cm^{-3}, a range common in plasma processing conditions.

Table 1. Simulation conditions with various α, β, γ, and probe diameters. Parameters *a–g* are given in Figure 1.

Type	Ring-Type Probe			Type	Bar-Type Probe		
Number	α ($\equiv d/a$)	β ($\equiv b/a$)	γ ($\equiv c/a$)	Number	α ($\equiv f/e$)	β ($\equiv h/e$)	γ ($\equiv g/e$)
#1	1	4.5	0.5	#8	4	4.5	0.5
#2	2	4.5	0.5	#9	6	4.5	0.5
#3	4	4.5	0.5	#10	10	4.5	0.5
#4	1	5.5	0.5	#11	4	5.5	0.5
#5	1	7.0	0.5	#12	4	7.0	0.5
#6	1	4.5	1.0	#13	4	4.5	1.0
#7	1	4.5	1.5	#14	4	4.5	1.5

3. Simulation Results and Discussion

To analyze the antenna characteristics of the two probes, Sections 3.1 and 3.2 include an investigation of the S_{21} spectrum in a vacuum condition (without plasma) with various geometrical parameters, such as α, β, γ, *a* for the RCP, and *e* for the BCP. Furthermore, for simple understanding, visualizations of the antenna configurations in the RCP and BCP are provided for each condition. Afterward, in plasma conditions, the S_{21} spectrum with various antenna distances is examined.

3.1. Ring-Type Cutoff Probe (RCP)

Figure 2a shows the S_{21} spectra with various α (normalized width of the detecting antenna) at fixed β and γ. An increase in α led to a slight elevation of the S_{21} level. This was because of an increase in the capacitive coupling between the radiating and detecting antenna. Furthermore, the S_{21} had several resonance frequencies at its extremes. Except for the lowest resonance frequency, with an increase in α, the resonance frequencies shifted toward lower values. When *a* increased, as shown in Figure 2d,g, the low-frequency shift became enlarged. The origin of the resonance frequencies will be analyzed later.

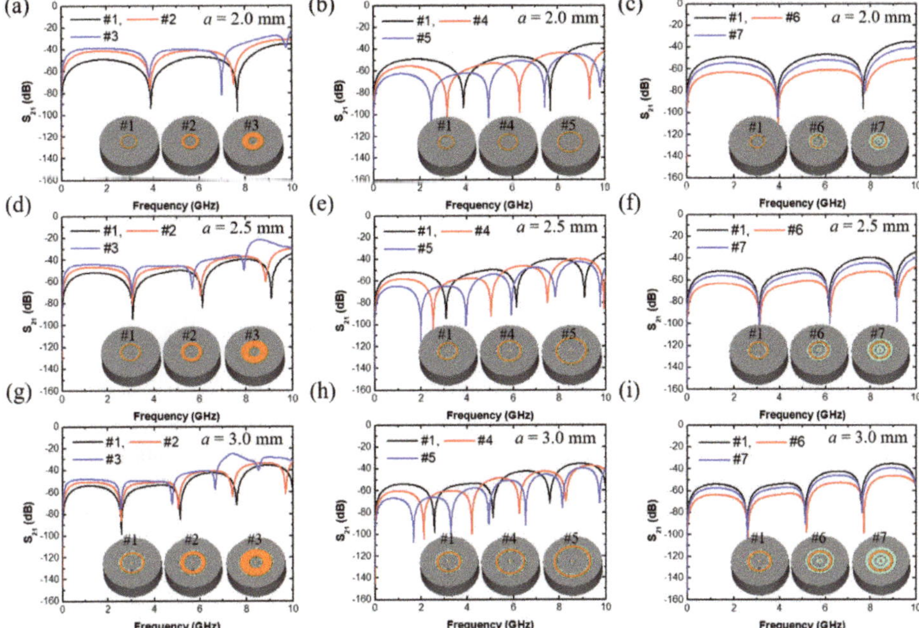

Figure 2. S_{21} spectra of the RCP from various (**a,d,g**) antenna widths, (**b,e,h**) antenna distances, and (**c,f,i**) dielectric thickness at various antenna diameters: (**a–c**) $a = 2.0$ mm, (**d–f**) $a = 2.5$ mm, and (**g–i**) $a = 3.0$ mm.

Figure 2b plots the S_{21} spectra with several β (normalized antenna distance), with results showing that the overall S_{21} signal level reduced with increasing β due to a decrease in the capacitive coupling between the antennae [23]. Besides, an increase in β led to a low frequency shift of the resonance frequencies. Compared with the effect of α, increasing β more clearly shifted the resonance frequencies toward lower values.

As shown in Figure 2c, an increase in γ (normalized dielectric thickness) did not change the resonance frequencies. The signal levels changed with γ, while there was no trend compared with the α and β effects. This might have resulted from a characteristic impedance mismatching between the radiation and detecting antennae, since they had no symmetry with each other. Impedance matching might deteriorate at $\gamma = 1.0$.

To figure out the origin of the resonance frequencies as shown in Figure 2, the electric field on the detecting antenna surface was examined, the direction of which was normal to the antenna surface (z-axis as shown in Figure 1). Figure 3 exhibits the normalized electric field as a function of the normalized antenna length ($\hat{\theta}$-axis as shown in Figure 1) at resonance (2.61, 5.17, and 7.71 GHz) and non-resonance (3.90 and 6.50 GHz) frequencies of the #6 condition at $a = 3.0$ mm. The simulation result showed that the oscillation amplitude at the resonances was larger than that at the non-resonances. Furthermore, the wavelength at the three resonances was one, two, and three times the antenna length. Based on these two facts, the resonance frequencies resulted from the SW resonances of the electric field on the detecting antenna [20,24]. Due to this SW resonance, electric field energy was strongly localized on the detecting antenna surface such that it could not propagate toward the Signal Out (Figure 1); the S_{21} value at the resonances, therefore, dramatically decreased, as shown in Figure 2.

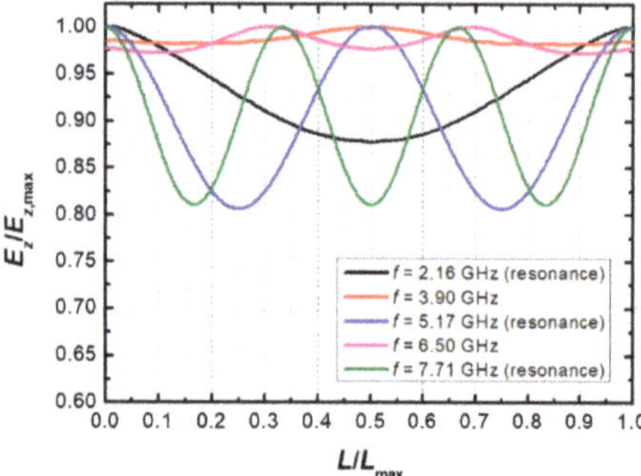

Figure 3. Normalized E_z-field distribution ($E_z/E_{z,max}$) on the detecting antenna of the RCP along normalized antenna length (L/L_{max}), where the z-direction is parallel to the normal direction of the antenna surface.

The RCP essentially featured resonance peaks induced by SW resonance in the simulated frequency ranges (<10 GHz), resulting from the length of the detecting antenna. This fact ultimately brought some negative aspects to the RCP, since the resonance peaks made finding the cutoff peak difficult in an S_{21} spectrum, as discussed below.

Figure 4 shows the S_{21} spectra of the RCP with various electron density values ranging from 1×10^9 to 1×10^{11} cm^{-3} with various β. In Figure 4a, an increase in β produced a high frequency shift of the cutoff frequency, which was the extreme in the S_{21} spectrum and marked by an arrow. Eventually, the cutoff frequency became equal to the plasma frequency (f_{pe} in Figure 4). This might have resulted from a larger rate of reduction in the capacitive coupling in the sheath (Path 1 in Figure 1c) compared to that in the plasma (Path 2 in Figure 1c) with increasing antenna distance [25]. In other words, at a short antenna distance, capacitive coupling in the sheath was dominant, so the effective electron density became smaller than the input electron density due to the sheath, where the electron density was zero.

This means that the cutoff frequency was lower than the plasma frequency. But at longer antenna distances, coupling in the plasma was dominant, and the effective electron density became the input electron density, and thereby, the cutoff frequency matched the plasma frequency [26]. Deeper analysis of the different coupling reduction trends between Path 1 and Path 2 in terms of antenna distance is beyond the scope of this paper, but will be discussed in detail in a later paper with rigorous theory.

The trend of cutoff frequency saturating to the plasma frequency at large antenna distances was the same as for other high density cases, as shown in Figure 4b,c. If measurement accuracy is defined as the discrepancy between the cutoff frequency (f_c) and the plasma frequency (f_{pe}) as $\frac{f_{pe}-f_c}{f_{pe}}$, the measurement accuracy of the RCP, therefore, strongly depended on the antenna distance; it was recommended that the distance be as large as possible. However, since the overall level of an S_{21} spectrum diminished as the antenna distance increased, there was a trade-off between signal-to-noise ratio and measurement accuracy. Additionally, the larger the antenna distance was, the more complex the spectrum shape became due to the SW resonance.

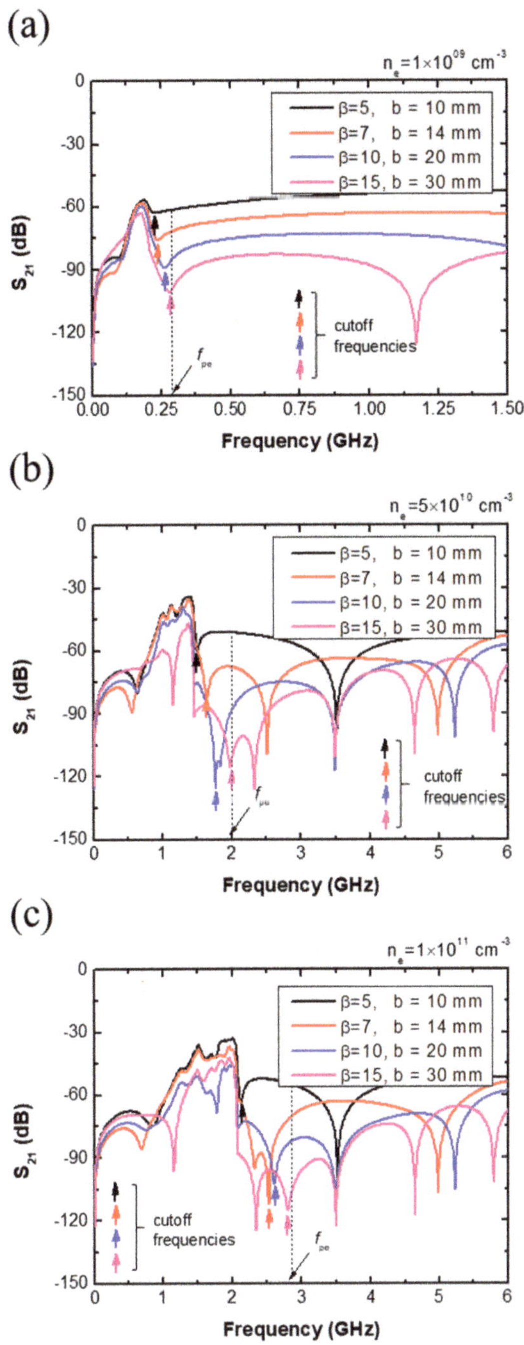

Figure 4. S_{21} spectra of the RCP from various normalized antenna distances (β) with electron densities of (**a**) $n_e = 1 \times 10^9$ cm^{-3}, (**b**) 5×10^{10} cm^{-3}, and (**c**) 1×10^{11} cm^{-3}.

3.2. Bar-Type Cutoff Probe (BCP)

Figure 5a shows the S_{21} spectra of the BCP with various α at fixed β and γ in a vacuum condition. An increase in α led to a slight increase of the S_{21} level, which was similar to the RCP case and for the same reason. Here, there were no significant resonance peaks in the S_{21} spectra from the BCP, except for case #3, because the length of the radiating and detecting antennae was much smaller than the RCP detecting antenna; the SW resonance took place at high frequencies beyond the interested range (<10 GHz). When the antenna length e increased, as shown in Figure 5d,g, a low frequency shift of the resonance frequency with low Q-factor took place near 8 GHz (cases #2 and #3), but this could be negligible. The signal level of the BCP, however, was lower than that of the RCP, by as much as 20 dB due to the small size of the BCP antennae.

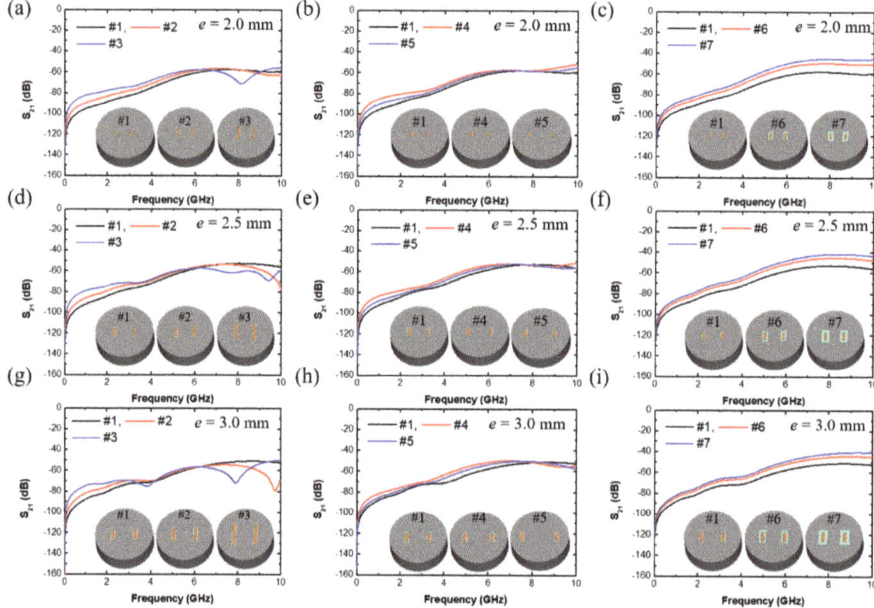

Figure 5. S_{21} spectra of the BCP from various (**a,d,g**) antenna widths, (**b,e,h**) antenna distances, and (**c,f,i**) dielectric thickness at various antenna lengths: (**a–c**) e = 2.0 mm, (**d–f**) e = 2.5 mm, and (**g–i**) e = 3.0 mm.

Figure 6 shows the S_{21} spectra of the BCP with various electron densities ranging from 1×10^9 cm^{-3} to 1×10^{11} cm^{-3} and various β, with the same conditions as in Figure 4. In Figure 6a, an increase in β produced the same results as the RCP case. With increasing antenna distance, the cutoff frequency saturated to the plasma frequency, which might have resulted from a reduction in capacitive coupling between the antennae for the same reason as in the RCP. Figure 6b,c show smooth spectral shapes, and it is therefore easy to determine the cutoff frequency in each spectrum. They also manifested the same trend as Figure 6a in terms of β dependence.

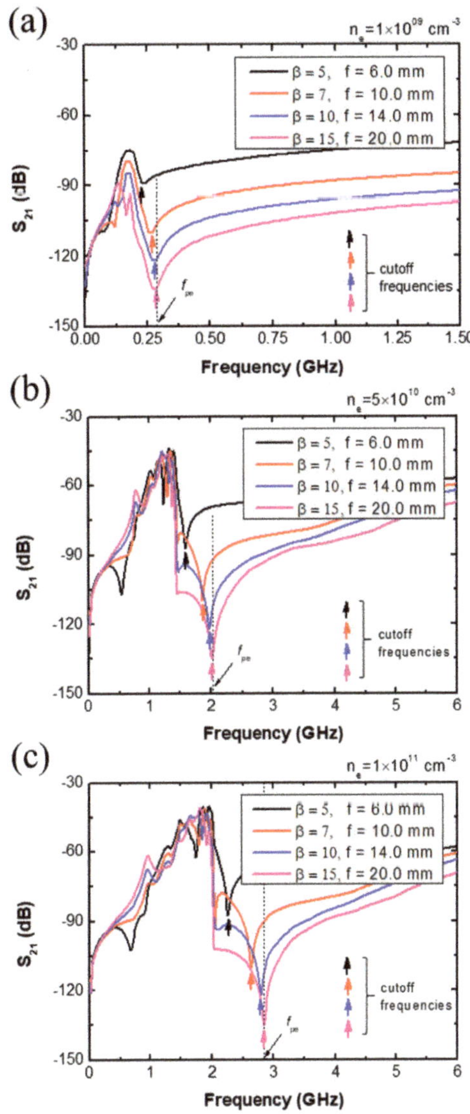

Figure 6. S_{21} spectra of the BCP from various normalized antenna distances (β) with electron densities of (**a**) $n_e = 1 \times 10^9$ cm^{-3}, (**b**) 5×10^{10} cm^{-3}, and (**c**) 1×10^{11} cm^{-3}.

It should be noted for the BCP here that the critical antenna distance (d_c), where the cutoff frequency matched the plasma frequency, was much lower than that in the RCP, at 14 mm compared to 30 mm as shown in Figure 7, which showed the ratio of the cutoff frequency to the plasma frequency by antenna distance. Additionally, the S_{21} spectrum of the BCP was much straighter, without any resonance peaks, than that of the RCP. This fact facilitated simple cutoff frequency monitoring by just measuring the minimum S_{21} value, which indicated the possibility for a simple plasma monitoring system based on the BCP. In summary, the BCP was more practical than the RCP in terms of miniaturization as well as electron density monitoring, since it was much easier to determine the cutoff frequency in the S_{21} spectrum of the BCP than that of the RCP.

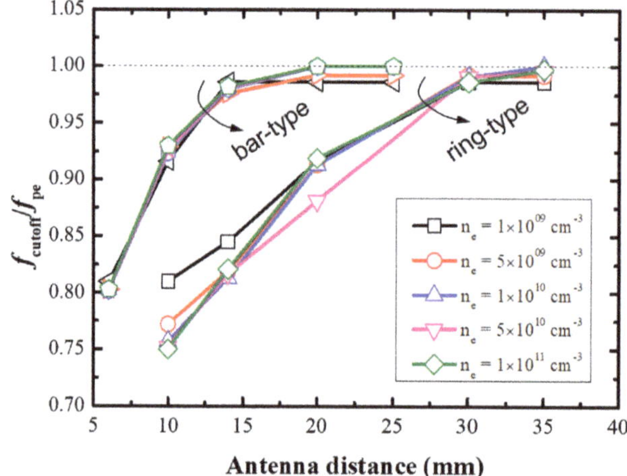

Figure 7. Normalized cutoff frequency ($f_{\text{cutoff}}/f_{\text{pe}}$) of the RCP and BCP with various antenna distances and input electron densities at $a = e = 2.0$ mm and $p = 100$ m Torr.

4. Conclusions

This paper investigated the basic properties of two types of planar cutoff probes, the ring-type and the bar-type planar cutoff probes, with various geometrical parameters as well as a plasma parameter through a commercial 3D electromagnetic simulation. Simulation results showed that the RCP had several resonance frequencies that originated from standing-wave resonance on the detecting antenna, while the BCP did not. Moreover, the signal level of both the RCP and BCP increased with increasing antenna size and decreased with increasing antenna distance between the radiating and detecting antennae because of elevated and diminished capacitive coupling, respectively. Among the studied parameters, antenna distance was found to be the key parameter highly related to the accuracy of both probes. As a result, the BCP was more practical than the RCP for use in non-invasive electron density measurement systems since it allowed for much easier determination of cutoff frequency in S_{21} spectra.

Author Contributions: Investigation, S.J.K.; resources, H.C.L., J.-H.K.; data curation, S.J.K., J.J.L., Y.S.L., and H.J.Y.; writing—original draft preparation, S.J.K.; writing—review and editing, S.J.Y. All authors have read and agreed to the published version of the manuscript.

Funding: This research was supported by the National Research Council of Science & Technology (NST) grant by the Korea government (MSIP) (No. CAP-17-02-NFRI), by the Korea Institute of Energy Technology Evaluation and Planning (KETEP) and the MOTIE of the Republic of Korea (No. 20172010105910), by Korea Institute for Advancement of Technology (KIAT) grant funded by the Korea Government (MOTIE) (P0008458, The Competency Development Program for Industry Specialist), by the Aerospace Low Observable Technology Laboratory Program of the Defense Acquisition Program Administration and the Agency for Defense Development of the Republic of Korea, and by Basic Science Research Program through the National Research Foundation of Korea(NRF) funded by the Ministry of Education (NRF-2020R1A6A1A03047771).

Conflicts of Interest: The authors declare no conflict of interest.

References

1.　Lieberman, M.A.; Lichtenberg, A.J. *Principles of Plasma Discahrges and Materials Processing*, 2nd ed.; Wiley & Sons. Inc.: Hobken, NJ, USA, 2005; pp. 1–22.

2.　Adamovich, I.V.; Baalrud, S.D.; Bogaerts, A.; Bruggeman, P.J.; Cappelli, M.; Colombo, V.; Czarnetzki, U.; Ebert, U.; Eden, J.G.; Favia, P.; et al. The 2017 Plasma Roadmap: Low temperature plasma science and technology. *J. Phys. D Appl. Phys.* **2017**, *50*, 323001. [CrossRef]

3. Umeda, S.; Nogi, K.; Shiraishi, D.; Kagoshima, A. Advanced Process Control Using Virtual Metrology to Cope With Etcher Condition Change. *IEEE Trans. Semicond. Manuf.* **2019**, *32*, 423–427. [CrossRef]

4. Roh, H.-J.; Ryu, S.; Jang, Y.; Kim, N.-K.; Jin, Y.; Park, S.; Kim, G.-H. Development of the Virtual Metrology for the Nitride Thickness in Multi-Layer Plasma-Enhanced Chemical Vapor Deposition Using Plasma-Information Variables. *IEEE Trans. Semicond. Manuf.* **2018**, *31*, 232–241. [CrossRef]

5. Ahn, J.-H.; Gu, J.-M.; Han, S.-S.; Hong, S.-J. Real-time In-situ Plasma Etch Process Monitoring for Sensor Based-Advanced Process Control. *J. Semicond. Technol. Sci.* **2011**, *11*, 1–5. [CrossRef]

6. Kim, S.J.; Lee, J.J.; Kim, D.W., Kim, J.-II.; You, S.J. A transmission line model of the cutoff probe. *Plasma Sources Sci. Technol.* **2019**, *28*, 055014. [CrossRef]

7. Engeln, R.; Klarenaar, B.L.M.; Guaitella, O. Foundations of optical diagnostics in low-temperature plasmas. *Plasma Sources Sci. Technol.* **2020**, *29*, 063001. [CrossRef]

8. You, K.H.; You, S.J.; Kim, D.W.; Na, B.K.; Seo, B.H.; Kim, J.H.; Seong, D.; Chang, H.-Y. Measurement of electron density using reactance cutoff probe. *Phys. Plasmas* **2016**, *23*, 053515. [CrossRef]

9. Kim, J.-H.; Choi, S.-C.; Shin, Y.-H.; Chung, K.-H. Wave cutoff method to measure absolute electron density in cold plasma. *Rev. Sci. Instruments* **2004**, *75*, 2706–2710. [CrossRef]

10. 3ds SIMULIA CST STUDIO SUITE. Available online: www.cst.com (accessed on 6 October 2020).

11. Liang, I.; Nakamura, J.; Sugai, H. Modeling Microwave Resonance of Curling Probe for Density Measurements in Reactive Plasmas. *Appl. Phys. Express* **2011**, *4*, 066101. [CrossRef]

12. Arshadi, A.; Brinkmann, R.P.; Hotta, M.; Nakamura, K. A simple and straightforward expression for curling probe electron density diagnosis in reactive plasmas. *Plasma Sources Sci. Technol.* **2017**, *26*, 045013. [CrossRef]

13. Arshadi, A.; Brinkmann, R.P. Analytical investigation of microwave resonances of a curling probe for low and high-pressure plasma diagnostics. *Plasma Sources Sci. Technol.* **2016**, *26*, 15011. [CrossRef]

14. Hotta, M.; Ogawa, D.; Nakamura, K.; Sugai, H. Real-time curling probe monitoring of dielectric layer deposited on plasma chamber wall. *Jpn. J. Appl. Phys.* **2018**, *57*, 046201. [CrossRef]

15. Ogawa, D.; Nakamura, K.; Sugai, H. A novel technique for in-situ simultaneous measurement of thickness of deposited film and electron density with two curling probes. *Plasma Sources Sci. Technol.* **2020**. [CrossRef]

16. Schulz, C.; Styrnoll, T.; Awakowicz, P.; Rolfes, I. The Planar Multipole Resonance Probe: Challenges and Prospects of a Planar Plasma Sensor. *IEEE Trans. Instrum. Meas.* **2014**, *64*, 857–864. [CrossRef]

17. Friedrichs, M.; Oberrath, J. The planar Multipole Resonance Probe: a functional analytic approach. *EPJ Tech. Instrum.* **2018**, *5*, 7. [CrossRef]

18. Kim, D.W.; You, S.J.; Kim, S.J.; Kim, J.-H.; Lee, J.Y.; Kang, W.S.; Hur, M.S. Planar cutoff probe for measuring the electron density of low-pressure plasmas. *Plasma Sources Sci. Technol.* **2019**, *28*, 015004. [CrossRef]

19. Yeom, H.J.; Kim, J.-H.; Choi, D.; Choi, E.S.; Yoon, M.Y.; Seong, D.-J.; You, S.J.; Lee, H.-C.; Hyo-Chang, L. Flat cutoff probe for real-time electron density measurement in industrial plasma processing. *Plasma Sources Sci. Technol.* **2020**, *29*, 035016. [CrossRef]

20. Kim, D.W.; You, S.J.; Kim, J.H.; Chang, H.Y.; Oh, W.Y. Computational comparative study of microwave probes for plasma density measurement. *Plasma Sources Sci. Technol.* **2016**, *25*, 35026. [CrossRef]

21. Kim, H.C.; Iza, F.; Yang, S.S.; Radmilović-Radjenović, M.; Lee, J.K. Particle and fluid simulations of low-temperature plasma discharges: benchmarks and kinetic effects. *J. Phys. D Appl. Phys.* **2005**, *38*, R283–R301. [CrossRef]

22. Verboncoeur, J.P. Particle simulation of plasmas: review and advances. *Plasma Phys. Control. Fusion* **2005**, *47*, A231–A260. [CrossRef]

23. Jackson, J.D. *Classical Electrodynamics*, 3rd ed.; Wiley & Sons. Inc.: Hobken, NJ, USA, 1999; p. 310.

24. Na, B.-K.; Kim, D.-W.; Kwon, J.-H.; Chang, H.-Y.; Kim, J.-H.; You, S.J. Computational characterization of cutoff probe system for the measurement of electron density. *Phys. Plasmas* **2012**, *19*, 53504. [CrossRef]

25. Walker, J.; Halliday, D.; Resnick, R. *Principles of Physics*, 10th ed.; John Wiley & Sons: Singapore, 2014; pp. 413–416.

26. Mehdizadeh, M. *Microwave/RF Applicators and Probes: for Material Heating, Sensing, and Plasma Generation*, 2nd ed.; Elsevier Inc.: Cambridge, MA, USA, 2015; p. 99.

Article

Transition from Microstrip Line to Ridge Empty Substrate Integrated Waveguide Based on the Equations of the Superellipse

David Herraiz [1], Héctor Esteban [1,*], Juan A. Martínez [2], Angel Belenguer [2], Santiago Cogollos [1], Vicente Nova [1] and Vicente E. Boria [1]

[1] Instituto de Telecomunicaciones y Aplicaciones Multimedia, Universitat Politècnica de València, Camino de Vera, s/n, 46022 Valencia, Spain; daherza@teleco.upv.es (D.H.); sancobo@dcom.upv.es (S.C.); vinogi@iteam.upv.es (V.N.); vboria@dcom.upv.es (V.E.B.)

[2] Departamento de Ingeniería Eléctrica, Electrónica, Automática y Comunicaciones, Universidad de Castilla-La Mancha, Escuela Politécnica de Cuenca, Campus Universitario, 16071 Cuenca, Spain; juanangel.martinez@uclm.es (J.A.M.); angel.belenguer@uclm.es (A.B.)

* Correspondence: hesteban@dcom.upv.es; Tel.: +34-619842369

Received: 22 October 2020; Accepted: 13 November 2020; Published: 16 November 2020

Abstract: In recent years, multiple technologies have been proposed with the aim of combining the characteristics of traditional planar and non-planar transmission lines. The first and most popular of these technologies is the Substrate Integrated Waveguide (SIW), where rows of metallic vias are mechanized in a printed circuit board (PCB). These vias, together with the top and bottom metal layers of the PCB, form a channel for the propagation of the electromagnetic fields, similar to that of a rectangular waveguide, but through a dielectric body, which increases the losses. To reduce these losses, the empty substrate integrated waveguide (ESIW) was recently proposed. In the ESIW, the dielectric is removed from the substrate, and this results in better performance (low profile and easy manufacturing as in SIW, but lower losses and better quality factor for resonators). Recently, to increase the operational bandwidth (monomode propagation) of the ESIW, the ridge ESIW (RESIW) and a transition from RESIW to microstrip line was proposed. In this work, a new and improved wideband transition from microstrip line (MS) to RESIW, with a dielectric taper based on the equations of the superellipse, is proposed. The new wideband transition presents simulated return losses in a back-to-back transition greater than 20 dB in an 87% fractional bandwidth, while in the previous transition the fractional bandwidth was 82%. This is an increment of 5%. In addition, the transition presents simulated return losses greater than 26 dB in an 84% fractional bandwidth. For validation purposes, a back-to-back configuration of the new transition was successfully manufactured and measured. The measured return loss is better than 14 dB with an insertion loss lower than 1 dB over the whole band.

Keywords: substrate integrated waveguide; ridge waveguide; tapering structure; broadband; microwave devices

---

1. Introduction

The growing need for small microwave devices with integration capabilities in planar technologies and high performance prompted the creation of the substrate integrated waveguide (SIW) [1]. In the quest for increasing the performance, the empty substrate integrated waveguide (ESIW) was developed [2]. In the ESIW, the dielectric is removed, the walls are metalized, and the electromagnetic wave is confined by a bottom and top metal covers. The removal of the dielectric allows a significant reduction of the losses, with the drawback of increasing the overall size. Consequently, the wave is propagated through the air. To validate this promising technology, some common microwave devices have been designed, manufactured, and measured. Some of these devices are filters [3,4], antennas [5–7], 90° hybrid directional couplers [8], circulators [9], and phase shifters [10].

All of the above devices need a transition to connect ESIW to planar technologies. Multiple transitions from microstrip line (ML) to ESIW have been designed with a different approach in the design, solving the matching problems and being functional despite the manufacturing tolerances [2,11–13]. However, these transitions have the disadvantage of being well matched in a limited frequency range (monomode propagation in ESIW), resulting in a fractional bandwidth (FBW) of approximately 42%. This fractional bandwidth may be insufficient for some broadband applications. Recently, a new microstrip line to ESIW transition that can achieve a higher fractional bandwidth (57%) has been proposed [14]. This transition is based on introducing a dielectric material inside the waveguide, which allows for a greater bandwidth, but with the drawback of increasing the losses. In this work, we aim to achieve larger fractional bandwidth than 57% without introducing dielectric and therefore increasing the losses.

In the ESIW, the fundamental mode TE_{10} is propagated in a similar way as in a standard rectangular waveguide. The frequency range in which this is the only mode that is propagating determines the usable bandwidth. With the intention of increasing the monomode bandwidth a ridge waveguide can be used. The ridge waveguide consists of a waveguide with one or more metallic ridges. These ridges produce a reduction in the frequency of the fundamental mode and a small increase in the frequency of the next higher order mode. Thus, a significantly larger monomode bandwidth is achieved. Therefore, a ridge waveguide with the same dimensions of a rectangular waveguide without ridge can work at lower frequencies. Conversely, a ridge waveguide can work at the same frequency as the conventional rectangular waveguide, but being smaller and less bulky.

This idea of a Ridge ESIW (RESIW) was first presented by Murad et al. [15], and a transition from microstrip line to a single RESIW (RESIW with only one ridge) that allows the use of most of the usable bandwidth was also proposed by Herraiz et al. [16]. This transition for a WR62 RESIW presents return losses greater than 20 dB in a fractional bandwidth (FBW) of 82%.

In this work, a thorough study of possible geometries of a ridge ESIW manufactured with standard Planar Circuit Board (PCB) machinery is performed with the purpose of assessing the pros and cons of using only one o more ridges, and/or using ridges with different cross sectional shapes, and selecting the best geometry with a compromise between monomode propagation bandwidth and ease of manufacturing. Next, a new wideband transition from microstrip line to single RESIW is proposed. This new transition is based in a wider ESIW section with a superelliptical dielectric taper, and it presents better return losses and a broader fractional bandwidth than the previous microstrip to single RESIW transition [16].

The work is structured as follows. Section 2 presents an exhaustive study of the possible geometries of the RESIW. Section 3 presents the transition and it is design procedure. The results of the simulated and measured prototype are presented in Section 4. Finally, the conclusions are discussed in Section 5.

2. RESIW cross Section Design

The ridge waveguides are a family of waveguides which are composed of one or several metallic ridges attached to the lower or upper walls of the waveguide [17]. The best known and studied ridge waveguides are the single and double ridge rectangular waveguides [18]. The metallic ridges allow a reduction of the cutoff frequency of the fundamental mode (TE_{10}), and an increase of the cutoff frequency of the first higher order mode (TE_{20}). Thus, a broader bandwidth can be achieved. In addition, the characteristic impedance of the fundamental mode (Z_0) is reduced [19].

In practice, by properly selecting the geometrical parameters of the single and the double ridge waveguides, similar bandwidths can be achieved with either one. Therefore, to reduce the complexity of the manufacture, the single RESIW is our focus of study. The use of the conventional manufacturing techniques of planar technologies allows the creation of this single RESIW simply by piling substrates of the same height. This leads to an easy manufacture, but limits the freedom in the design of the ridge dimensions, as shown in Figure 1. The height of the ridge (h_r) is limited by the height of the substrate (h_s) and can only be a multiple of this height ($h_r = n\,h_s$, where n is the number of ridge layers). The only parameter of the ridge transversal section that can be chosen freely is the ridge width w_r.

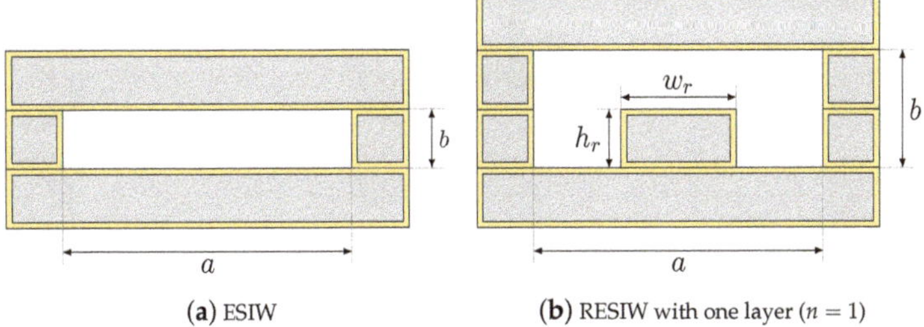

(**a**) ESIW   (**b**) RESIW with one layer ($n = 1$)

Figure 1. Cross sectional view of the ESIW and the ridge ESIW with one layer.

Appendix A shows that the one layer rectangular ridge ESIW proves to be the best compromise between maximizing the bandwidth and minimizing the related manufacturing cost and volume. Therefore, one-layer rectangular ridge ESIW is used in this work, where the waveguide width is $a = 15.7988$ mm (Standard waveguide WR-62) and the ridge width that maximizes the bandwidth is $w_r = 3.848$ mm.

3. Transition Design

The 3D view of the proposed topology for an improved wideband transition from microstrip line to one layer RESIW is shown in Figure 2a. The detailed layout with dimensions of the central (Figure 3a) and ridge layers (Figure 3b) of this transition is shown in Figure 3.

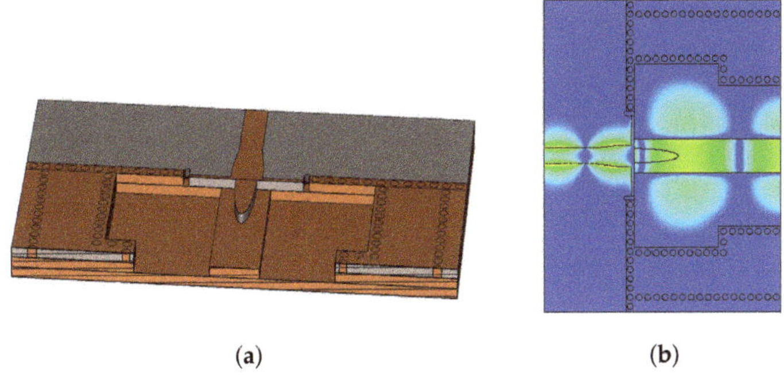

(**a**)   (**b**)

Figure 2. Structure of the proposed transition from microstripo to RESIW (**a**); and simulated field distribution for the central frequency (14 GHz) (**b**).

The transition is based in four elements in order to couple the fields and match the impedances of the input and output transmission lines. Since there is a large difference between the impedances of the microtrip (low impedance) and the RESIW (high impedance), the first element consists on a linear taper at the end of the microstrip line that gradually modifies the impedance of the microstrip line, thus easing the impedance matching with the RESIW. The second element of the transition is a widened RESIW section, placed at the beginning of the RESIW. This section is controlled with the parameters a_c (length) and b_c (width). This widening properly decreases the impedance of the RESIW, so that the impedance matching with the microstrip is easier. This widening is manufactured in the central and in the ridge layer (see Figure 3a,b). The third element of the transition is an inductive iris of width w_{ti}, mechanized with rows of metallized via holes, that controls the coupling between the widened microstrip line and the widened RESIW. These three elements were already present in the previous version of the MS-RESIW transition [16] (see Figure 4a). In that previous transition, the matching was completed with the fourth element of the transition, which was a widening of the ridge at the beginning of the RESIW [16]. In this work, instead of using a widening of the ridge, the ridge has no widening and a dielectric taper is used in the central layer in order to better match the discontinuity between dielectric and air. The use of a dielectric taper, never used before in a transition from MS to RESIW, is inspired by the dielectric taper used in the transition from microstrip line to ESIW presented in [2] and improved in [11] (see Figure 4b). However, and conversely to what happens in the transition from MS to ESIW, the exponential dielectric taper provides poor results in the case of a transition from MS to RESIW. For that reason, in this work, we propose the use of the equations of the superellipse [20] for shaping the dielectric taper. This curve allows to explore an endless set of possible transition profiles, using just a few number of parameters and finding the optimum parameter values that provide with an excellent performance.

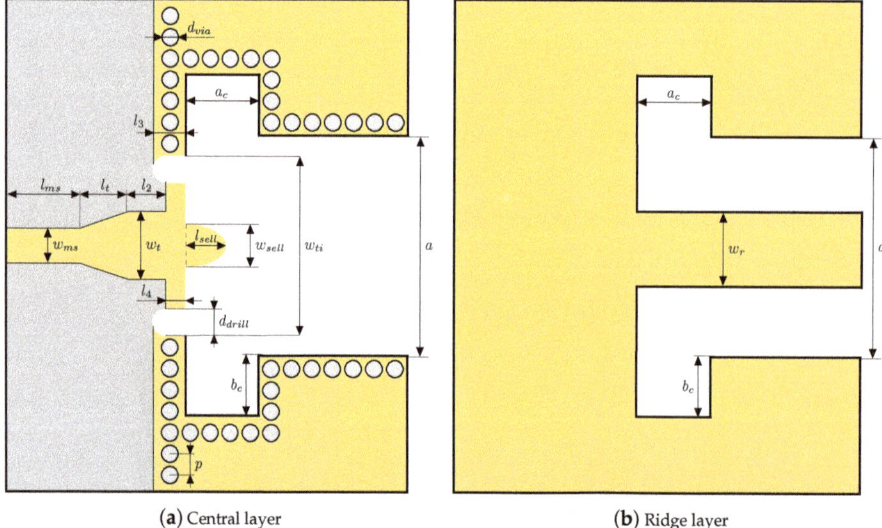

(a) Central layer   (b) Ridge layer

Figure 3. Layout of the microstrip line to RESIW transition proposed in this work. White represents empty spaces, gray represents dielectric body, and dark yellow represents copper surfaces.

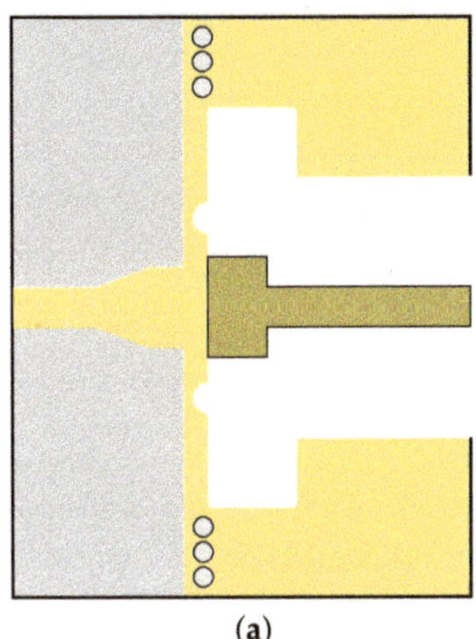
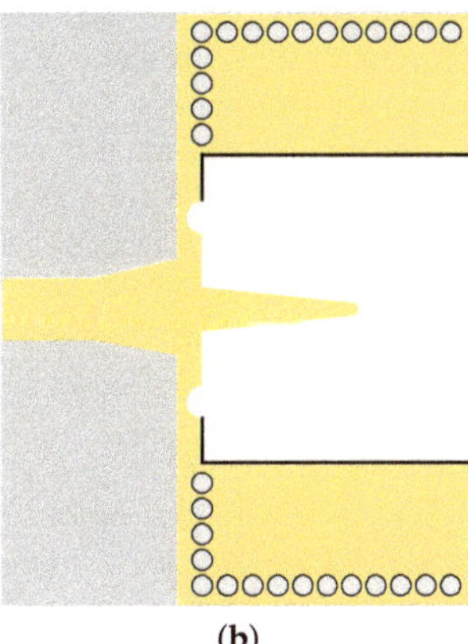

(a)   (b)

Figure 4. Previous transitions that inspired the transition presented in this work: (**a**) transition from MS to RESIW with widening of the ridge [16]; and (**b**) transition from MS to ESIW with exponential dielectric taper [11].

The equations of the superellipse have been used in several fields of the engineering, especially in architecture [21]. In the microwave field, they were used in [20] for defining a tapering impedance matching profile between two rectangular waveguides of different widths. The superellipse was first discussed in 1818 by Gabriel Lamé. It consists in a generalization of the equations of the ellipse, so that the ellipse is a particular case of the superellipse. The Cartesian equation of superellipse is,

$$\left(\frac{x}{a}\right)^m + \left(\frac{y}{b}\right)^m = 1 \tag{1}$$

where b and a are the semiaxes of the superellipse and m is the parameter controlling the overall shape of the ellipse. By properly selecting the value of m, a rectellipse ($m = 4$), an ellipse ($m = 2$), a straight line ($m = 1$), an astroid ($m = 2/3$), or other shapes can be obtained. The most known and studied values of m and its shapes are shown in Figure 5.

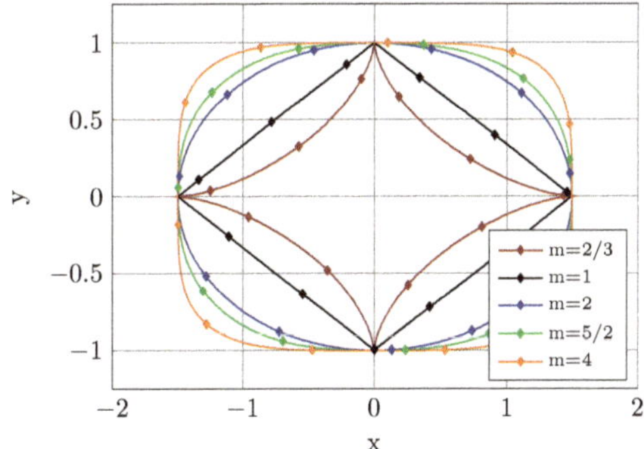

Figure 5. Superellipses with semiaxes $a = 1.5$ and $b = 1$ and different values for m.

With the aim of providing a better control over the shape of the superellipse, the equations can be further generalized as,

$$\left(\frac{x}{a}\right)^p + \left(\frac{y}{b}\right)^q = 1 \tag{2}$$

where p and q are real parameters that control the curvature of the superellipse in each dimension separately. Values of p less than one provide an inward curve with the y-axis. On the other hand, values of p greater than one provide an outward curve with the y-axis. q controls the curvature in the x-axis. Extreme values for p or q (i.e., $p, q = 0$ or $p, q \to \infty$) produce a nose sharp curve or a blunt ellipse.

As shown in Figure 3, the idea is to create a dielectric taper whose semiaxes are l_{sell} and $w_{sell}/2$ and find the best possible values of p and q that provide the maximum impedance matching. Thus, Equation (2) is particularized for creating a dielectric taper that follows the following profile,

$$\left(\frac{x}{l_{sell}}\right)^p + \left(\frac{y}{\frac{w_{sell}}{2}}\right)^q = 1 \tag{3}$$

To accelerate the process of designing the proposed transition, it is convenient to find good initial values for the design parameters. Table 1 proposes expressions that provide with a good initial point that significantly reduces the computational cost of the optimization process. The expressions for the initial values of l_t, l_2, and w_t are extracted from [16]. The other expressions were obtained empirically, after designing several transitions at different frequencies and with different substrates. The value of λ_g for computing the initial expressions refers to the wavelength in the widened section of RESIW of width $a + 2b_c$.

Table 1. Initial expressions for the design parameters of the new MS to RESIW transition.

w_{ti}	w_t	l_t	a_c	b_c	w_{sell}	l_{sell}	l_2	p	q
$0.7a$	$0.2w_{ti}$	$\frac{\lambda_{ms}(f_0)}{4}$	$\frac{\lambda_g(f_0)}{2}$	$\frac{a}{6}$	$0.4w_r$	$\frac{\lambda_g(f_0)}{4}$	$\frac{l_t}{3}$	2	2

4. Results

To validate the transition, a back-to-back transition from a 50 Ω microstrip line to a RESIW with one ridge layer was designed, manufactured, and measured. A Rogers 4003C substrate with $h_r = 0.813$ mm, permittivity $\epsilon_r = 3.55$, and metal thickness $t = 35$ μm was used. The transition presents the same width of the standard WR-62 rectangular waveguide ($a = 15.7988$ mm). The expressions in Table 1 were applied to obtain initial values for the design parameters. Since the initial values were good enough, the optimum values were obtained after a quick and simple optimization process with the Neder-Mead Simplex algorithm of CST and the goal of maximizing the return loss. Table 2 shows the initial values of the design parameters and the final optimum values. It also shows the dimensions that are fixed and that do not need to be optimized.

Table 2. Dimensions of the designed MS to RESIW transition with Rogers 4003C substrate (in mm).

Fixed Parameters			Design Parameters								
			Initial	Final		Initial	Final		Initial	Final
p, l_3	1										
d_{via}	0.7	w_{ti}	11.059	10.014	l_{sell}	4.518	4.996	l_t	3.213	4.428
w_r	3.848	b_c	2.633	2.394	l_2	1.071	1.491	p	2.00	2.008
d_{drill}	0.5	a_c	11.836	9.629	w_t	2.211	2.446	q	2.00	2.016
w_{ms}	1.813	w_{sell}	1.528	1.635	l_4	0.3	0.329			

Figure 6 compares the simulated (with CST) reflection coefficient of the new transition presented here and the previous version presented in [16]. To assess the validity of the initial values of the design parameters, the reflection coefficient with these initial values is also plotted. It can be observed that the initial values provide a minimum return loss of about 12 dB in the whole band, which proves that the initial vales of Table 1 are a very good approximation to the final optimum values. Comparing the optimized transition proposed here and the previous transition presented in [16], it can be observed that the fractional bandwidth (calculated as the frequency range in which the return losses are greater than 20 dB divided by the central frequency, 14 GHz) is 82% with the previous transition of

Herraiz et al. [16] and 87% with the new transition, which supposes an increment of 5%. It can also be observed that the optimized new transition presents a return losses value greater than 26 dB from 8.7 to 20.5 GHz, which is a very good matching in a large bandwidth, thus proving its good performance in terms of simulated results.

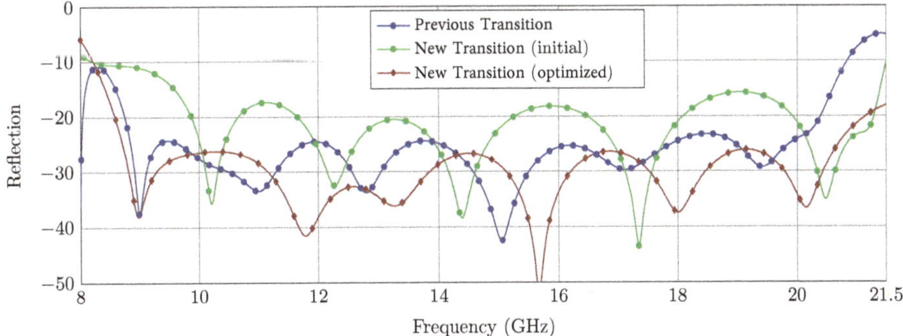

Figure 6. Return loss comparison between previous [16] and new version of the back-to-back transition from microstrip to RESIW. For the new transition, results of the design with initial values of the parameters and of the design with final optimized values are plotted.

The simulated insertion losses of the previous [16] and new transition are compared in Figure 7. Lossy metal and lossy dielectric are considered in the simulation with CST. It can be observed that the insertion losses of the new transition are lower than 0.6 dB in the whole band, whereas the insertions losses of the previous transition are greater than 1.2 dB.

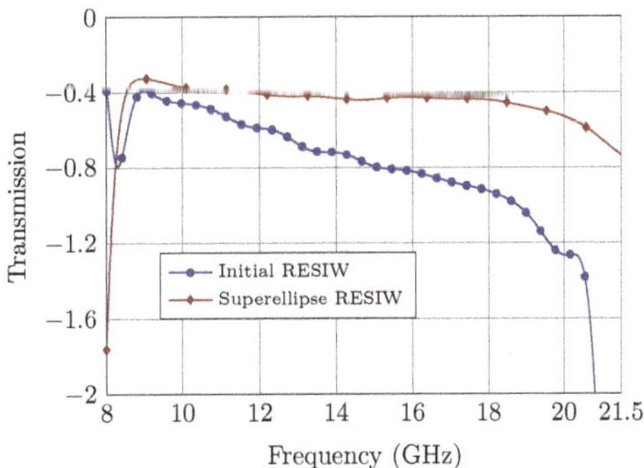

Figure 7. Insertion loss comparison between previous [16] and new back-to-back transition from microstrip to RESIW.

Therefore, the simulated results of the new version of the transition are better in terms of both insertion and return losses than the previous version of Herraiz et al. [16].

Figure 8 compares the return losses of the the back-to-back transition from microstrip to ESIW without ridge [11] and the new back-to-back transition from microstrip to RESIW presented here. There is an increment of 103% in the fractional bandwidth (return losses greater than 20 dB) with the new transition when compared with the transition from MS to ESIW without ridge, whose fractional bandwidth is 42%.

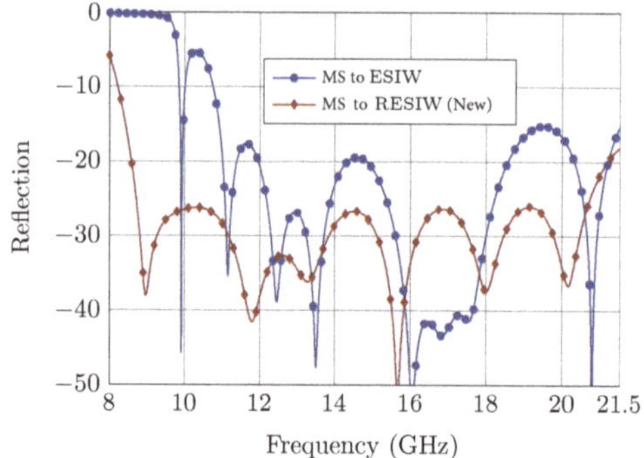

Figure 8. Return loss comparison between back-to-back transition from microstrip to ESIW [11] and this work.

The manufactured back-to-back prototype with the final design parameters of Table 2 is shown in Figure 9. Standard planar circuit manufacturing techniques (drilling, milling, and electrodeposition) were used to manufacture this prototype. Figure 10 compares the simulated and measured scattering parameters of the back-to-back transition prototype. The differences between measurement and simulation are mainly due to manufacturing tolerances and the effect of surface roughness on electrodeposited metallic surfaces. For the same cut, the tolerance of the laser beam is only 2 μm and for nonconsecutive cut is approximately 50 μm. The tolerance in the alignment of the layers is the same as of nonconsecutive cuts. The measurements include the effect of the transitions from coaxial to microstrip, which can be estimated in approximately 1 dB of additional insertion losses, and possibly some mismatch. Discounting the effect of the coaxial to microstrip connectors, the measured insertion loss of the back-to-back prototype is smaller than 1 dB, and the measured return loss is greater that 14.5 dB between 9 and 20.66 GHz, which corresponds to a fractional bandwidth of 83%. In [16], the return loss was 11 dB and the insertion loss was 1.5 dB between 8 and 20.5 GHz, which corresponds to a fractional bandwidth of 87% for a back-to-back transition with the same waveguide characteristics.

In comparison, the measured return and insertion losses of the new version of the transition are better than the previous one reported in [16].

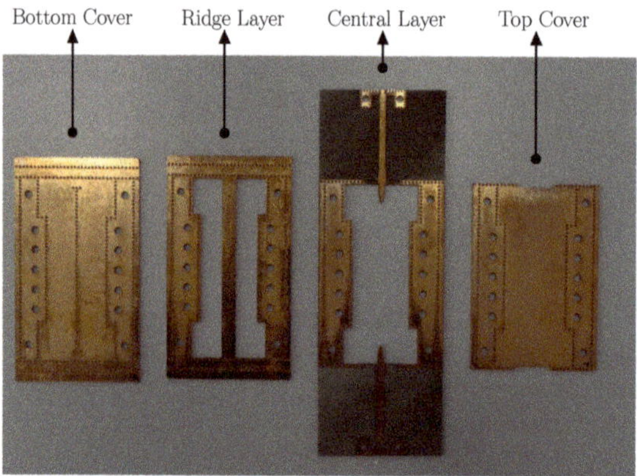

Figure 9. Back-to-back manufactured prototype.

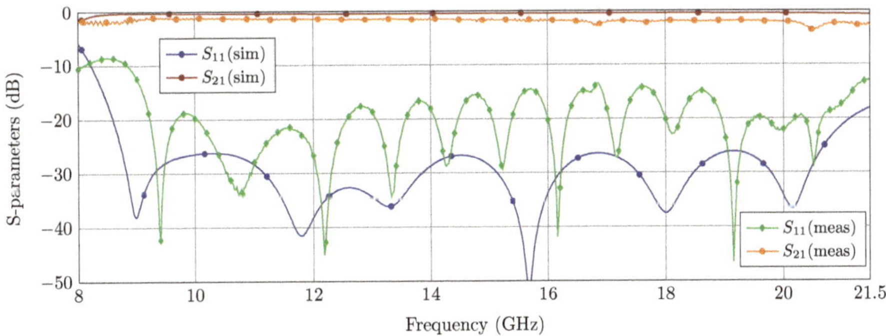

Figure 10. Comparison of measurements and simulations of the back-to-back transition prototype.

5. Conclusions

In this work, a novel transition from MS to RESIW based on the equations of the superellipse is proposed. A design procedure is presented and initial values for the design parameters are provided. The simulated reflection coefficient of a back-to-back transition is less than 20 dB in an 87% fractional bandwidth and less than 26 dB in an 84% fractional bandwidth. The measured return losses are greater than 14.5 dB and the insertion losses are lower than 1 dB in a 83% fractional bandwidth. The new wideband transition presents improved (simulated and measured) responses, with better return losses and lower insertion losses, when compared with the previous MS to RESIW transition.

Author Contributions: Conceptualization, D.H., S.C., and H.E.; methodology, D.H. and H.E.; software, D.H.; validation, D.H., V.N., and J.A.M.; formal analysis, D.H.; investigation, D.H. and H.E.; resources, A.B. and V.E.B.; writing—original draft preparation, D.H.; writing—review and editing, H.E.; supervision, A.B., V.E.B., and H.E.; project administration, A.B. and V.E.B.; and funding acquisition, A.B. and V.E.B. All authors have read and agreed to the published version of the manuscript.

Funding: This research was funded by Ministerio de Ciencia e Innovación, Spanish Government, under Research Projects PID2019-103982RB-C44 and PID2019-103982RB-C41.

Conflicts of Interest: The authors declare no conflict of interest. The founding sponsors had no role in the design of the study; in the collection, analyses, or interpretation of data; in the writing of the manuscript, and in the decision to publish the results.

Abbreviations

The following abbreviations are used in this manuscript:

SIW	Substrate Integrated Waveguide
PCB	Printed Circuit Board
ESIW	Empty Substrate Integrated Waveguide
RESIW	Ridge Empty Substrate Integrated Waveguide
FBW	Fractional Bandwidth
MS	Microstrip Line
CST	Computer Simulation Technology
WR	Rectangular Waveguide
RL	Return Loss

Appendix A. Study of Possible Geometries of a RESIW Crosssection

In this Appendix, an exhaustive study of the possible geometries of the RESIW is presented. Figure A1 shows the RESIWs transversal section that have been studied.

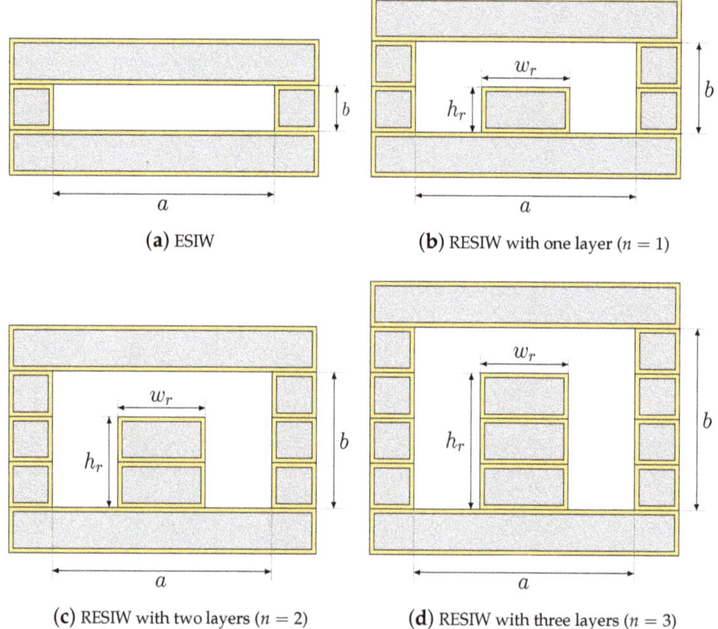

Figure A1. Cross sectional view of the ESIW and the ridge ESIW with different number of layers.

In the ridge waveguide, the cutoff frequencies of the fundamental mode TE_{10} (f_{c10}) and the first higher order mode TE_{20} (f_{c20}) can be obtained by solving the following transcendental equations [22],

$$\cot\left(\frac{2\pi f_{c_{10}}\left(\frac{a-w_r}{2}\right)}{c}\right) - \frac{b}{d}\tan\left(\frac{\pi w_r f_{c_{10}}}{c}\right) - \frac{B}{Y_{01}} = 0 \tag{A1}$$

$$\cot\left(\frac{2\pi f_{c_{20}}\left(\frac{a-w_r}{2}\right)}{c}\right) + \frac{b}{d}\cot\left(\frac{\pi w_r f_{c_{20}}}{c}\right) - \frac{B}{Y_{01}} = 0 \tag{A2}$$

where c is the light velocity; a and b are the width and height of the waveguide, respectively; and $d = b - h_r$ is the distance between the top ridge layer and top wall of waveguide (see Figure 1). The term B/Y_{01} is the susceptance that models the gap, and can be approximated as [17]

$$\frac{B}{Y_{01}} \approx 4\left(\frac{b}{d}\right)\left(\frac{a f_c}{c}\right)\ln\csc\left(\frac{\pi d}{2b}\right) \tag{A3}$$

where f_c is f_{c10} for solving Equation (A1) and f_{c20} for solving Equation (A2).

Equations (A1) and (A2) can be used to obtain f_{c10} and f_{c20} as a function of the width of the ridge, w_r, which is the only design parameter in a RESIW that we can use in order to maximize the usable bandwidth ($f_{c20} - f_{c10}$). Figure A2 shows both cutoff frequencies f_{c10} and f_{c20} as a function of w_r for a RESIW manufactured with Rogers 4003C substrates of height $h = 0.813$ mm, permittivity $\epsilon_r = 3.66$ and width $a = 15.7988$ mm. The cutoff frequencies were computed both analytically with Equations (A1) and (A2) and numerically with the commercial software Computer Simulation Technology (CST). The results are shown for a RESIW with one, two and three ridge layers ($n = 1, 2$ and 3), and they are compared with the cutoff frequencies of an ESIW (no ridge layers).

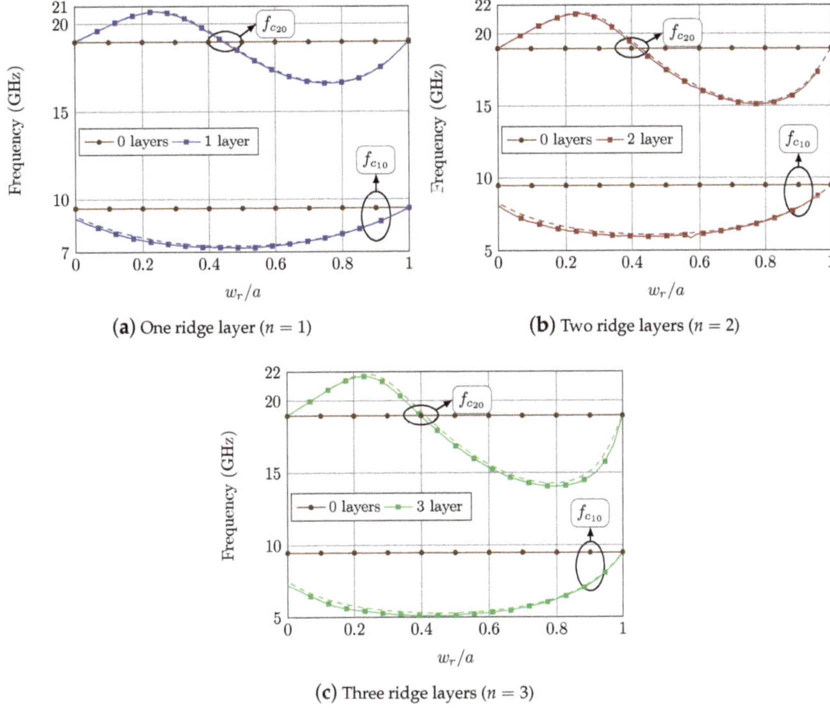

(a) One ridge layer ($n = 1$)

(b) Two ridge layers ($n = 2$)

(c) Three ridge layers ($n = 3$)

Figure A2. Cutoff frequencies of the fundamental and first higher order mode of a ridge ESIW with one, two and three ridge layers as a function of the ridge width w_r. Comparison between analytical (dashed lines) and simulated (continuous lines) results and comparison with an ESIW without any ridge layer.

It can be observed, in the first place, that there is a very good agreement between analytical and simulated results, which validates the accuracy of Equations (A1) and (A2). Besides, it can be concluded that the higher bandwidth (maximum distance between f_{c10} and f_{c20}) is achieved with $w_r/a \simeq 0.28$. Using this value of w_r as a starting point, and after optimization, optimum values of w_r that maximize the bandwidth were obtained for the the RESIW with one, two, and three ridge layers, and the values of the cutoff frequencies for these optimum values of w_r are shown in Table A1.

Table A1. Cutoff frequencies and bandwidth of the RESIW as the number of ridge layers (n) increases.

n	f_{10}(GHz)	f_{20}(GHz)	BW(GHz)	BW Increase
0	9.480	18.970	9.490	-
1	7.530	20.660	13.130	38%
2	6.227	21.352	15.125	15%
3	5.420	21.692	16.272	7%

The results depicted in Table A1 show that the bandwidth increases with the number of ridge layers. Using one ridge layer increases the bandwidth by 38% compared with the ESIW with no ridge layer. However, using two ridge layers supposes an increase in the bandwidth of only a 15% when compared with one ridge layer. Using three ridge layers increases the bandwidth by 7% compared with two ridge layers. Taking into account that increasing the number of ridge layers increases the volume, weight, cost, and manufacturing complexity, for most wideband applications, the best compromise between bandwidth and ease of manufacturing will be the use of one ridge layer.

To test whether the usable bandwidth of the RESIW with one ridge layer can still be improved by using other unconventional geometries, the trapezoidal and two rectangular ridge geometries shown in Figure A3 were also studied.

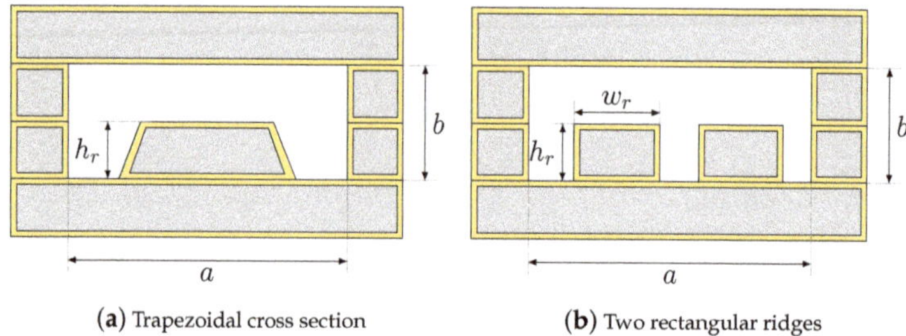

(**a**) Trapezoidal cross section (**b**) Two rectangular ridges

Figure A3. Cross sectional view of the one layer ridge ESIW with unconventional geometries.

Table A2 shows the cutoff frequencies and the absolute bandwidth of these geometries compared with the rectangular cross section single layer ridge. It can be observed that a slightly higher bandwidth can be achieved with the trapezoidal geometry, but the improvement is so small that it is not worth complicating the manufacturing process.

Table A2. Cutoff frequencies and bandwidth for RESIW with on ridge layer and different ridge geometries.

Ridge Geometry	f_{10}**(GHz)**	f_{20}**(GHz)**	BW**(GHz)**
One rectangular ridge	7.530	20.660	13.130
One trapezoidal ridge	7.710	20.860	13.150
Two rectangular ridges	7.650	19.950	12.300

References

1. Deslandes, D.; Wu, K. Integrated microstrip and rectangular waveguide in planar form. *IEEE Microw. Wirel. Compon. Lett.* **2001**, *11*, 68–70. [CrossRef]
2. Belenguer, A.; Esteban, H.; Boria, V. Novel Empty Substrate Integrated Waveguide for High-Performance Microwave Integrated Circuits. *IEEE Trans. Microw. Theory Tech.* **2014**, *62*, 832–839. [CrossRef]
3. Martinez, J.A.; de Dios, J.J.; Belenguer, A.; Esteban, H.; Boria, V.E. Integration of a Very High Quality Factor Filter in Empty Substrate-Integrated Waveguide at Q-Band. *IEEE Microw. Wirel. Compon. Lett.* **2018**, *28*, 503–505. [CrossRef]
4. Belenguer, A.; Esteban, H.; Borja, A.L.; Boria, V.E. Empty SIW technologies: A major step toward realizing low-cost and low-loss microwave circuits. *IEEE Microw. Mag.* **2019**, *20*, 24–45. [CrossRef]
5. Mateo, J.; Torres, A.; Belenguer, A.; Borja, A. Highly Efficient and Well-Matched Empty Substrate Integrated Waveguide H-plane Horn Antenna. *IEEE Antennas Wirel. Propag. Lett.* **2016**, *15*, 1510–1513. [CrossRef]
6. Miralles, E.; Belenguer, A.; Mateo, J.; Torres, A.; Esteban, H.; Borja, A.; Boria, V. Slotted ESIW antenna with high efficiency for a MIMO radar sensor. *Radio Sci.* **2018**, *53*, 605–610. [CrossRef]
7. Khan, Z.U.; Alomainy, A.; Loh, T.H. Empty Substrate Integrated Waveguide Planar Slot Antenna Array for 5G Wireless Systems. In Proceedings of the 2019 IEEE International Symposium on Antennas and Propagation and USNC-URSI Radio Science Meeting, Atlanta, GA, USA, 7–12 July 2019; pp. 1417–1418.
8. Fernandez, M.D.; Ballesteros, J.A.; Belenguer, A. Design of a hybrid directional coupler in empty substrate integrated waveguide (ESIW). *IEEE Microw. Wirel. Compon. Lett.* **2015**, *25*, 796–798. [CrossRef]
9. Martinez, L.; Laur, V.; Borja, A.L.; Quéffélec, P.; Belenguer, A. Low Loss Ferrite Y-Junction Circulator Based on Empty Substrate Integrated Coaxial Line at Ku-Band. *IEEE Access* **2019**, *7*, 104789–104796. [CrossRef]

10. Peng, H.; Xia, X.; Ovidiu Tatu, S.; Xu, K.D.; Dong, J.; Yang, T. Broadband phase shifters using comprehensive compensation method. *Microw. Opt. Technol. Lett.* **2017**, *59*, 766–770. [CrossRef]
11. Esteban, H.; Belenguer, A.; Sánchez, J.R.; Bachiller, C.; Boria, V.E. Improved low reflection transition from microstrip line to empty substrate-integrated waveguide. *IEEE Microw. Wirel. Compon. Lett.* **2017**, *27*, 685–687. [CrossRef]
12. Liu, Z.; Xu, J.; Wang, W. Wideband Transition From Microstrip Line-to-Empty Substrate-Integrated Waveguide Without Sharp Dielectric Taper. *IEEE Microw. Wirel. Compon. Lett.* **2019**, *29*, 20–22. [CrossRef]
13. Belenguer, A.; Ballesteros, J.A.; Fernandez, M.D.; González, H.E.; Boria, V.E. Versatile, Error Tolerant, and Easy to Manufacture Through-Wire Microstrip-to-ESIW Transition. *IEEE Trans. Microw. Theory Tech.* **2020**, *68*, 2243–2250. [CrossRef]
14. Liu, Z.; Wang, W.; Sun, D.; Deng, J.Y. Wideband microstrip-to-air-filled substrate integrated waveguide transition with controllable band-pass performance. *Microw. Opt. Technol. Lett.* **2020**, *62*, 3458–3462. [CrossRef]
15. Murad, N.; Lancaster, M.; Wang, Y.; Ke, M. Micromachined rectangular coaxial line to ridge waveguide transition. In Proceedings of the 2009 IEEE 10th Annual Wireless and Microwave Technology Conference, Clearwater, FL, USA, 20–21 April 2009; pp. 1–5.
16. Herraiz, D.; Esteban, H.; Martínez, J.A.; Belenguer, A.; Boria, V. Microstrip to Ridge Empty Substrate-Integrated Waveguide Transition for Broadband Microwave Applications. *IEEE Microw. Wirel. Compon. Lett.* **2020**, *30*, 257–260. [CrossRef]
17. Marcuvitz, N. *Waveguide Handbook*; MacGraw Hill: London, UK, 1951.
18. Helszajn, J. *Ridge Waveguides and Passive MicrowaveComponents*; Number 49 of IET Electromagnetic Waves Series; IET Digital Library: London, UK, 2000.
19. Cohn, S.B. Properties of ridge wave guide. *Proc. IRE* **1947**, *35*, 783–788. [CrossRef]
20. Cogollos, S.; Vague, J.; Boria, V.E.; Martínez, J.D. Novel planar and waveguide implementations of impedance matching networks based on tapered lines using generalized superellipses. *IEEE Trans. Microw. Theory Tech.* **2018**, *66*, 1874–1884. [CrossRef]
21. Gardner, M. *Mathematical Carnival*; American Mathematical Society: Providence, RI, USA, 1965
22. Hopfer, S. The design of ridged waveguides. *IRE Trans. Microw. Theory Tech.* **1955**, *3*, 20–29. [CrossRef]

Publisher's Note: MDPI stays neutral with regard to jurisdictional claims in published maps and institutional affiliations.

MDPI

St. Alban-Anlage 66

4052 Basel

Switzerland

Tel. +41 61 683 77 34

Fax +41 61 302 89 18

www.mdpi.com

Applied Sciences Editorial Office

E-mail: applsci@mdpi.com

www.mdpi.com/journal/applsci